Theory and Interpretation of Magnetic Resonance Spectra

Theory and Interpretation of Magnetic Resonance Spectra

W.T.Dixon
Department of Chemistry, Bedford College
University of London, Regent's Park
London N.W.1

PLENUM PRESS · London and New York · 1972

Plenum Publishing Company Ltd.
Davis House
8 Scrubs Lane
Harlesden
London NW10 6SE
Tel. 01-969 4727

U.S. Edition published by
Plenum Publishing Corporation
227 West 17th Street
New York New York 10011

Softcover reprint of the hardcover 1st edition 1972

ISBN-13: 978-1-4684-7861-7 e-ISBN-13: 978-1-4684-7859-4
DOI: 10.1007/978-1-4684-7859-4

Library of Congress Catalog Card Number: 73-180922

Adlard & Son Ltd., Bartholomew Press, Dorking

Preface

It is amazing how much information can be gleaned from a magnetic resonance spectrum by one who knows. That series of lines on chart paper may conceal anything from energies of activation and spin densities, to conformations and differentiation of isomers. In order to be able to deduce such things about the structure and properties of molecules in a sample, it is necessary to be familiar with the underlying principles, and to arrive at that state of understanding is not easy.

This book was conceived and written in an attempt to clarify what is necessary theoretical equipment for anyone wishing to extract the maximum information from a magnetic resonance spectrum. It is also written for those who will find a fascination and great satisfaction in the way this subject, which involves so many sides of modern physics, holds together.

It seems to the author, from experience, that the difficulty of getting to grips with the theory of magnetism and magnetic resonance is two-sided. On the one hand one has forgotten, or never really known, the principles of electromagnetism on which it is based, and on the other, detailed analysis of the spectra requires a certain facility with the operator techniques of quantum mechanics. In both cases the principle difficulty appears to be unfamiliarity, so the chief aim in this book will be to introduce the enquirer to the relevent language in a reasonably connected fashion.

To this end the book is divided into two main parts. In the first three chapters the basic physics and quantum mechanics are covered, and in the others we consider first the analysis and then the interpretation of magnetic resonance spectra.

The exact choice of material is, of course, a subjective one, but it is hoped that any student who perseveres with the text will find himself able to set up and solve most of the problems he comes across in specific applications of magnetic resonance. This is one of the two main objects which the author has in mind—to provide the student with enough background to tackle actual problems. The other is to excite his interest and curiosity.

Where the text differs from previous ones is that it is more or less self-contained, i.e. the formulae and physical concepts used are all developed from first principles in the first part of the book.

Another novel feature is the way it is divided up to bring out the essential unity of nuclear and electron paramagnetic resonance, for the principles of nuclear magnetic resonance are the same as those of bulk magnetism and of electron spin resonance.

The author owes much to previous works which illustrate the interesting nature of the subject, especially those by Slichter; by Pople, Bernstein and Schneider; and by Carrington and McLachlan. He would also like to thank professor G. H. Williams for his advice and encouragement, to Mr Paul Ashworth for running some of the ESR spectra, and to Mr John Wiffen for Figure 7-4.

August 1971 *W.T.D.*

Contents

Chapter 1

Introduction and Basic Theory for a Particle in a Field

1.1. INTRODUCTION

An interesting definition of science is that it is a systematized study. This implies that a scientist's work is divided into two parts: first, he makes observations and secondly, he arranges them into some sort of order. Modern science, and in particular, physical science, can be regarded as one of the possible branches of development of "science", and its practical (i.e. observational) and theoretical (systematizing) methods have gradually become standardized. These methods are respectively the use of experiments to reproduce favourable conditions for observing a given phenomenon, and the assigning of numbers to certain aspects of observations, in order to facilitate the collecting and ordering (i.e. the interpretation) of corresponding "facts".

Attempts at systematizing these numerical facts have led to the construction of algebraic theoiies which can be used to show relationships between phenomena, which are often unconnected at first sight, and also to present the data from a vast number of experiments in as condensed a form as possible.

In this book we shall be mainly concerned with relating magnetic resonance spectra with each other and with other experimental systems, in terms of electromagnetic theory and what we can call "mechanics".

Before embarking on this project it is of value to remember that a theory consists of not only a quantitative part (i.e. a mathematical structure) but also a qualitative part, invented for use in discussing and thinking about phenomena. It is the qualitative part, consisting of concepts rather closely tied to our subjective impression of the world, which gives meaning to the algebraic symbols (i.e. the language) of the theory, and which is of interest to the more practical scientist, who may not have the time, inclination or aptitude to learn all of the mathematical structure. We shall try to act as a translator between the mathematical and conceptual languages wherever it is practicable.

At this point it is useful to describe what we mean by the term "magnetic resonance".

Sub-atomic particles, such as electrons and nuclei, often have associated with them a magnetic moment, said to arise from an intrinsic angular momentum which in turn corresponds to the classical idea of "spin". In a magnetic field the magnetic moment of a particle can have certain orientations, each associated with its own potential energy. Now for a given field, only a few levels of magnetic energy are possible for a particle, so that to induce transitions between them we have to use radiation of a particular frequency, given by the equation $\Delta E = h\nu$ (ΔE = energy difference) (ν = frequency). It is the net absorption of radiation of a "preferred" frequency, when a particle is in a magnetic field, which is called "magnetic resonance".

In atomic and molecular systems, magnetic forces are very much smaller than electric ones, and this means that we have to go much deeper into the theory in discussing magnetic properties than, say, in getting the general idea of electronic energy levels or of vibrating/rotating molecules. We shall therefore have to look at some general ideas in electromagnetism and quantum theory before going on to examine specific effects in magnetic resonance. We shall examine (a) how "spin" arises, and how it interacts with fields; (b) how far large scale concepts can be carried over to sub-atomic phenomena and how they have to be modified (i.e. by quantum theory); (c) how a collection of spins ought to behave (implying the use of statistical mechanics).

1.2. ELECTROSTATICS AND MAGNETOSTATICS

The philosophy behind classical mechanics was that of "cause and effect", so that a change in the state of motion of a body was said to arise from the action of a force whose magnitude and direction is given by the rate of change of its momentum (mass × velocity). Forces between bodies arise because of something they contain, for example, they are attracted together by the force of gravity "because they contain a certain amount of mass".

When bodies contain electricity, produced perhaps by rubbing insulators together, from electrolytic cells or from dynamos, they exert forces on each other and the phenomena which arise can be rationalized by postulating three types of charge; positive, negative and neutral. In macroscopic systems we need to use only two of these, thinking of a neutral body as simply containing equal quantities of positive and negative charge.

Charges are measured by the forces they exert upon each other and the laws governing these forces are nicely summed up in an equation which states Coulomb's law:

$$\bar{F}_{12}=C(q_1q_2/r_{12}^2)(\bar{r}_{12}/r_{12}) \tag{1.1}$$

Where $\bar{F}_{12}$ is the force acting on the charge q_2, $\bar{r}_{12}$ is the vector distance $q_1 \rightarrow q_2$, r_{12} its magnitude, and $(\bar{r}_{12}/r_{12})$ its direction. C is a constant which defines the magnitude of our unit charge.

For gaussian or c.g.s. units

$$C=1$$

For S.I. or M.K.S. units

$$C=\frac{1}{4\pi\epsilon_0}$$

This equation tells us the rule "like charges repel, unlike charges attract" and applies to point charges *in vacuo*. The two other main laws of electrostatics are:

(i) *Conservation of Charge*

The total charge associated with a "closed" system remains constant with time.

e.g. Ionization in solution, $NaCl \rightleftharpoons Na^+ + Cl^-$.

e.g. Creation/annihilation of an electron–positron pair, $h\nu \rightleftharpoons e^- + e^+$.

(ii) *Invariance of Charge*

The magnitude of a charge is not affected by the relative motion of an observer.

e.g. The charges on isotopic nuclei are identical in spite of the very different motions of the constituent protons (as seen in their atomic spectra).

e.g. The exact electrical neutrality of the elements.

Quantization of Charge

From Milliken's experiments or from Faraday's laws of electrolysis and the atomic theory, it appears that electricity is quantized so that all charges are integral multiples of that on the electron.

One consequence of these empirical laws is that we can write down the electrostatic forces between systems of point charges, irrespective of whether they are moving, simply by adding the vectors given by Eq. (1.1) for each pair of charges. However, in atomic and molecular physics we can usually only specify the energy exactly, rather than the coordinates of the particles making up a system, so it is convenient to convert the "inverse-square law" Eq. (1.1) into potential energy form. To achieve this, suppose we take a unit charge experiencing a force $\bar{E}$, components E_x, E_y, E_z, in the x, y, z directions, from a point $P(xyz)$ to $Q(x+dx, y+dy, z+dz)$

high V

↓ force on positive charge

low V

change in potential energy = work done against the force

i.e.

$$dV = -(E_x dx + E_y dy + E_z dz) = -\bar{E}.d\bar{s}$$

hence $E_x = -\dfrac{\partial V}{\partial x}$ etc., or $\bar{E} = -\nabla V = -\text{grad}\ (V)$.

Fields

If a unit charge experiences a force $\bar{E}$ as in the above example, we take this as the definition of the strength of the field of force at that point. This field arises from some distribution of charges around the point and the concept is useful because we only have to know the *resultant* force on the unit charge in order to calculate its motion, and not necessarily any details of the charge distribution. The force on a charge q in a field $\bar{E}$ is simply $q\bar{E}$.

To give an example, the field of force due to a charge q is $q\bar{r}/r^3$ (in gaussian units), at a distance $\bar{r}$ (a vector field). The corresponding potential field is q/r (a scalar field), assuming that our zero of energy is at infinite separation.

We can use the idea of a field to reformulate Coulomb's law Eq. (1.1): consider a charge q at the centre of a sphere, radius r.

The field at the surface is $q\bar{r}/r^3$, at a point $\bar{r}$, so that we find the following relationship on differentiating:

$$\frac{\partial E_x}{\partial x} + \frac{\partial E_y}{\partial y} + \frac{\partial E_z}{\partial z} = \nabla.\bar{E} = \text{div}\ \bar{E} = 3q/r^3\ (r = \text{constant})$$

$$= 4\pi q/(4\pi r^3/3) = 4\pi(q/\text{volume})$$

i.e.

$$\nabla\bar{E} = \text{div}\ \bar{E} = 4\pi\rho \qquad (1.2)$$

where ρ is the charge density.

It is easy to extend this proof to the general case.

In S.I. units this formula is:

$$\text{div}\ \bar{E}=\rho/\epsilon_0 \tag{1.2a}$$

The theory of magnetism can be developed along exactly similar lines to the above treatment of electrostatics, the only difference being that there are no free magnetic poles—only dipoles. This means that the equation containing the inverse square law for magnetism, corresponding to Eq. (1.2), will be

$$\nabla.\bar{B}=\text{div}\ \bar{B}=0 \tag{1.3}$$

where $\bar{B}$ is the magnetic field strength at a point.

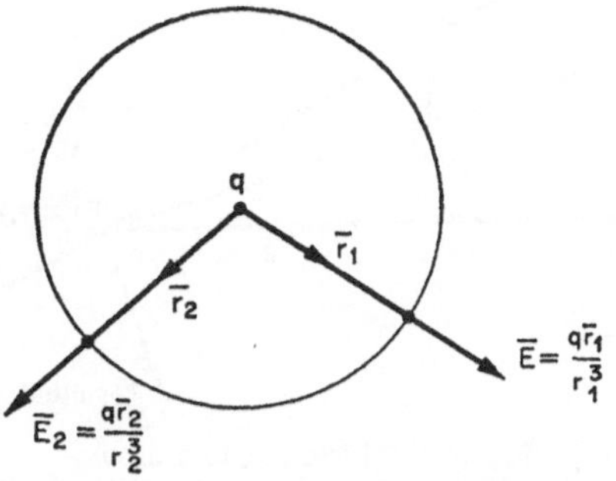

Figure 1-1. Field on a sphere due to a charge at its centre.

The force on a hypothetical pole "m" is $m\bar{B}$ and its field at a distance $\bar{r}$ is

$$\begin{cases} \mu_0 \dfrac{m\bar{r}}{4\pi r^3} \text{ in S.I. units} \\ \dfrac{m\bar{r}}{r^3} \quad \text{in c.g.s. units} \end{cases}$$

These two differential Eqs. (1.2) and (1.3), derived from the inverse square laws of electricity and magnetism are the first two of Maxwell's equations for free space.

1.3. INTERACTIONS OF DIPOLES AND QUADRUPOLES WITH FIELDS

In most problems in electromagnetism there are two main stages involved, i.e. first the field at given points has to be calculated, and secondly, the interaction of the system of particles with this field is found. To deal with the first part it is generally easier to calculate the potential field at a point because this is a scalar quantity. The corresponding force (vector) field can then be found by differentiation.

Field Due to a "Small" Dipole

As an example we shall find the field due to a dipole at distances large compared with its physical dimensions.

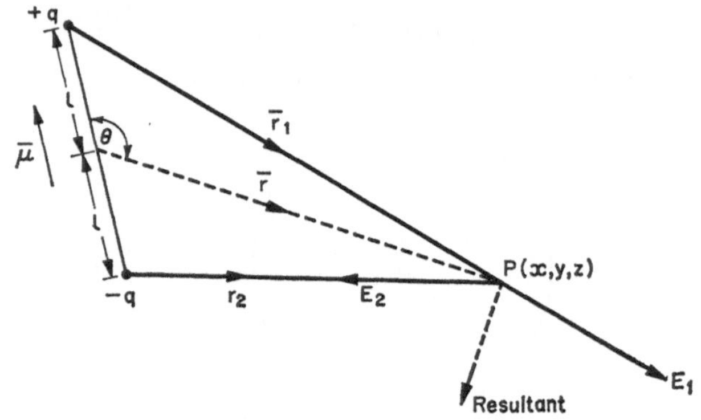

Figure 1-2. Field due to a dipole.

Our dipole consists of two charges (or poles), $+q$ and $-q$, distance $2l$ apart. We shall assume that $l \ll r$.

Potential at P is $+q/r_1 - q/r_2 = +q(r_2 - r_1)/r_2 r_1$ (c.g.s. units).

Since l is small, we have the approximate relation:

$$r_2 = r_1 + 2l \cos\theta$$

so defining the dipole moment $\mu = 2ml$, the potential at P is

$$V = \mu \cos\theta / r^2 = \bar{\mu} . \bar{r} / r^3 \tag{1.4}$$

The electric field in the x direction is found by differentiating with respect to x:

i.e. $$E_x = -\frac{\partial}{\partial x}(\mu_x x + \mu_y y + \mu_z z)/r^3 = -\mu_x/r^3 + 3x(\bar{\mu}.\bar{r})/r^5$$

i.e. $$\bar{E} = -\bar{\mu}/r^3 + 3(\bar{\mu}.\bar{r})\bar{r}/r^5 \tag{1.5}$$

E is the resultant of E_1 and E_2.

This equation applies for the fields due to either electric or magnetic dipoles. In S.I. units the right-hand side has to be multiplied by $1/4\pi\epsilon_0$ in the electrostatic case, and by $\mu_0/4\pi$ in the magnetic case.

Motion and Energy of a Dipole in a Field

Let us consider the situation when there is a field which we can consider homogeneous over the space occupied by a dipole. For simplicity let the direction of the field be that of the z axis, also let l_{xy} be the projection of the length of the dipole on to the xy plane. We shall look at the magnetic case.

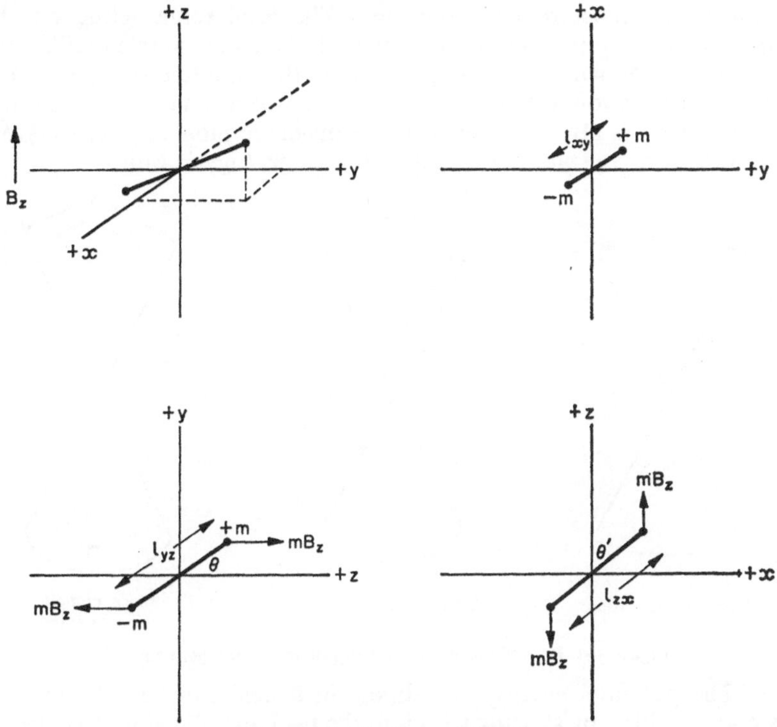

Figure 1-3. Dipole in a Field B_z.

Now torque is the rate of change of angular momentum so that the torque about z axis is zero (see Figure 1-3) i.e.

$$dM_z/dt=0$$

Torque about y axis (clockwise looking from $-y$ to $+y$) is $-2\ mB_z(l_{zx}/2)\sin\theta$, i.e.

$$dM_y/dt=-\mu_x B_z$$

Similarly the clockwise couple about the x axis is

$$dM_x/dt=+mB_z l_{yz}\sin\theta=+\mu_y B_z$$

These three results can be gathered together in a single vector equation i.e.

$$d\bar{M}/dt = \bar{\mu} \wedge B \tag{1.6}$$

where generally

$$dM_x/dt = \mu_y B_z - \mu_z B_y, \text{ etc.} \qquad \text{(see Appendix)}$$

These equations show us that the angular momentum about the applied field must remain constant. The total force acting on the dipole is zero so its motion must therefore be either simple oscillations in a plane, or some sort of precession, the simplest being called a Larmor precession which occurs when the total angular momentum of the dipole is always parallel to the magnetic moment, i.e. $\bar{\mu} = \gamma \bar{M}$, "$\gamma$" is then a constant and is called the magnetogyric ratio.

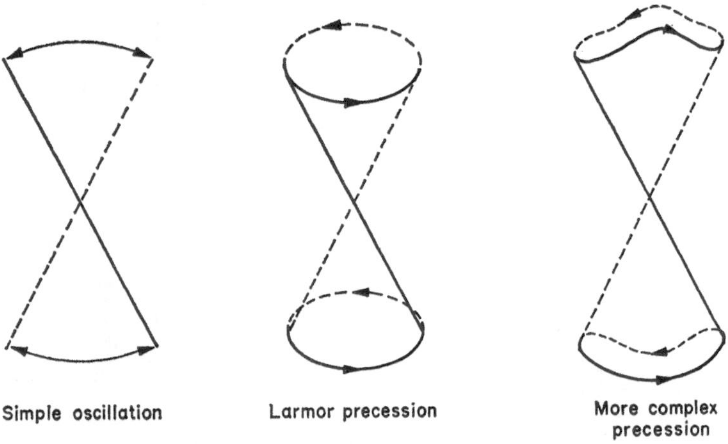

Figure 1-4. Possible motions of a dipole in a homogeneous field.

The potential energy of a dipole in a field can be calculated in two ways, either by starting with it in the field and allowing the orientation to change, or by bringing it into the field at a constant angle. We shall use the second method since it is less common and perhaps more instructive:

For simplicity we imagine taking the dipole from zero field into the final (homogeneous) field always keeping it at the same angle to the lines of force (see Figure 1-5)

force on the dipole in the position shown $= (E_1 - E_2)q$

total work done in taking it from zero to final field will be:

$$V = \Sigma(E_2 - E_1)(21 \cos \theta)q$$

$$= -E_{\text{final}}\mu \cos \theta$$

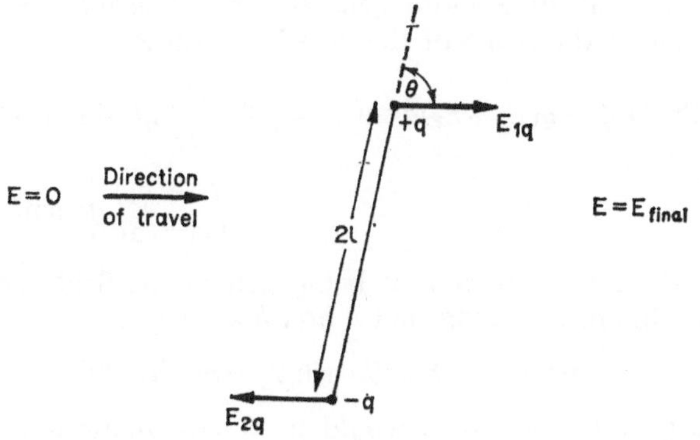

Figure 1-5.

i.e.

$$V=-\bar{E}.\bar{\mu}=-(\mu_x E_x+\mu_y E_y+\mu_z E_z) \tag{1.7}$$

We can now find an expression for the interaction energy V_{12} of two point dipoles μ_1, μ_2, using Eqs. (1.5) and (1.7) i.e.

$$V_{12}=\bar{\mu}_1.\bar{\mu}_2/r^3-3(\bar{\mu}_1.\bar{r})(\bar{\mu}_2.\bar{r})/r^5 \tag{1.8}$$

where $\bar{r}$ is the line joining the two dipoles.

Interaction of a Quadrupole with a Field

Just as we can form a dipole from two equal and opposite charges, so we can form a quadrupole from two equal and opposite dipoles (see Figure 1-6).

To simplify the problem we take the plane of the quadrupole to be parallel to the lines of force, this is equivalent to considering one

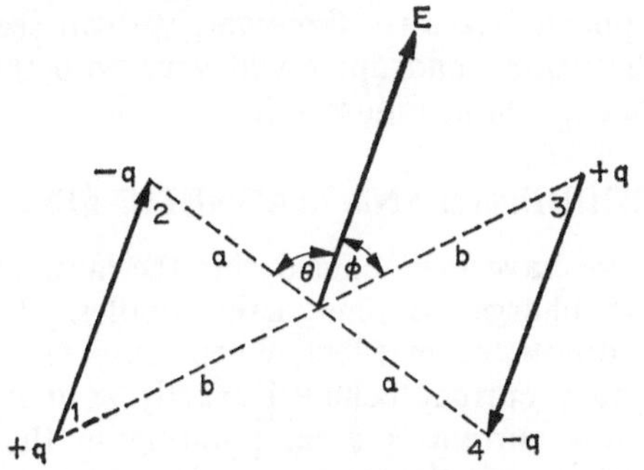

Figure 1-6. Quadrupole in a field.

of the projections in a more general case. Torque about an axis perpendicular to the plane of the parallelogram is

$$q(E_3-E_1)b \sin \phi - q(E_4-E_2)a \sin \theta = \frac{(E_3-E_1)}{(z_3-z_1)} qb \sin\phi\, (z_3-z_1) - \frac{(E_4-E_2)}{(z_4-z_2)} qa \sin\theta\, (z_4-z_2)$$

where we take the z direction as being that of the field. After some algebra we obtain, assuming that a and b are small:

$$\text{torque} = q\partial E/\partial z(b^2 \sin 2\phi - a^2 \sin 2\theta) \qquad (1.9)$$

In the general case there would be three equations like this for the three different projections.

Figure 1-7. Quadrupole in field due to a point charge.

The main point of interest at this stage is that a quadrupole does not interact noticeably with a homogeneous field but does experience a torque in a field gradient. If we wished we could get the potential energy in a similar way to that described for a dipole and this leads to a rather complicated result. However, we can see that the stable "static" orientation of a quadrupole will vary with the field gradient, a simple example is given in Figure 1-7.

1.4. STEADY CURRENTS AND MAGNETIC FIELDS

Up till now we have been considering the laws applying to either systems of static charges or magnetic dipoles. Classical electromagnetism was completed, in effect, when Ampere discovered that a small coil carrying a current behaved exactly as if it were a magnet whose moment was current × area, pointing in the same direction as the vector representing the area (i.e. along the normal). We can write this

$$\bar{\mu}=i\Delta\bar{S} \tag{1.10}$$

One implication of this equation is that we could, if we wished, abandon the idea of a magnetic pole altogether, and simply replace magnetic dipoles, where they occur, by appropriate combinations of "small coils". We, however, shall continue to use the concept of a magnet, since the mathematics is often greatly simplified when we replace a circulation of charge by the corresponding magnetic moment. An important consequence of Eq. (1.10) is that electric and magnetic fields are really parts of a single phenomenon, the electromagnetic field, as demonstrated by the other two of Maxwell's equations, which we shall now derive in a simple way.

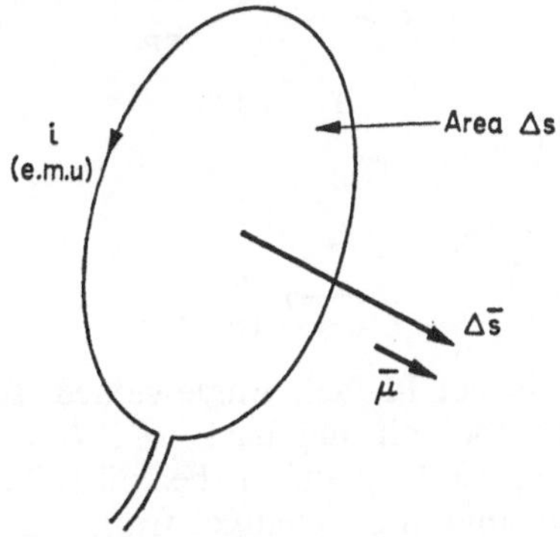

Figure 1-8.

First, we can use Eq. (1.10) to get expressions for the magnetic effects of currents, e.g. those of a finite coil, by imagining the circuit as being made up of a network of infinitesimal coils (see Figure 1-9). The potential at P due to the loop ΔS is

$$\mu \cos \theta/r^2=i\Delta S \cos \theta/r^2=id\omega$$

where $d\omega$ is the solid angle subtended at P.

The magnetic potential at P due to the whole coil is then:

$$\phi=\Sigma id\omega=i\omega \tag{1.11}$$

where ω is the solid angle subtended by the whole coil at P.

In S.I. units potential energy field at P is given by

$$V=i\Delta S \cos \theta \frac{\mu_0}{4\pi r^2}=\frac{\mu_0}{4\pi} i\omega$$

so defining $\phi=(1/\mu_0)V$, and $\bar{H}=(1/\mu_0)\bar{B}$, the magnetic potential ϕ is

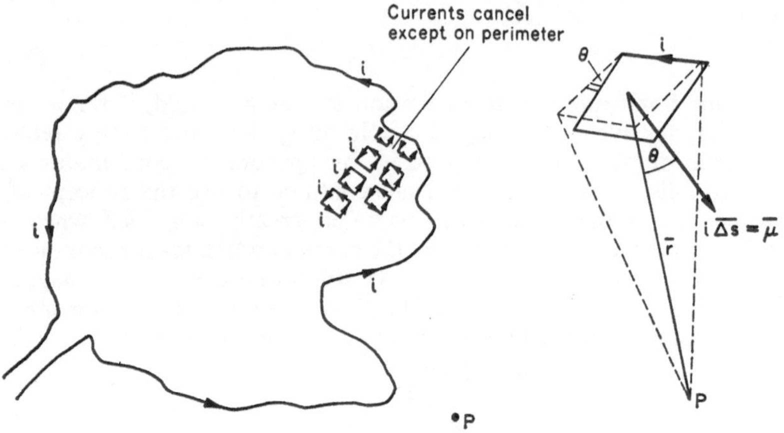

Figure 1-9.

given by

$$\phi = i\frac{\omega}{4\pi} \tag{1.11a}$$

The potential at P is not in fact, single-valued, for suppose that we take a path through the coil and back to P (see Figure 1-10). The change in potential up to the plane of the coil is $2\pi i - \omega i$. On passing through the coil the solid angle changes from $+2\pi$ to -2π, this corresponds to a change in potential of $4\pi i$; the angle then increases to its maximum in the negative sense, say ω'; then to zero at the side of the coil and finally back to ω at P. When we add up all the changes in potential we get zero, because the situation we have described corresponds to taking a charge through an electrical dipole "shell".

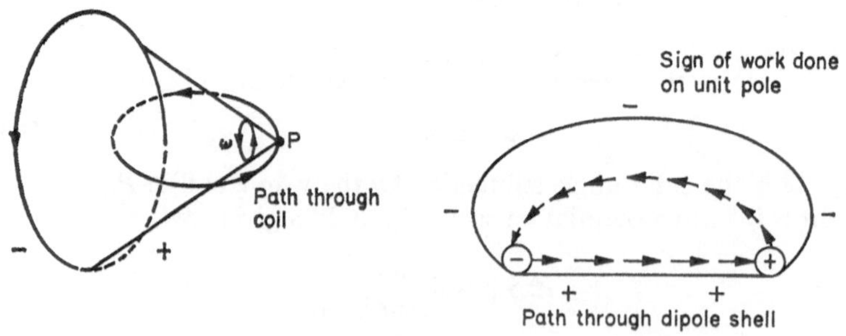

Figure 1-10.

The lines of force threading the coil are, however, continuous and do not reverse as they would have to if there really was a magnetic dipole shell, implying that no work is done in going from $\omega=2\pi$ to $\omega=-2\pi$, (via shortest route).

The change in potential on going around the path is therefore $+4\pi i$ and the idea of a scalar potential at a point is no longer useful, i.e. for the magnetic effects of currents. The result of all this is that we must make the assumption that there are no lines of force in one of our elementary magnetic dipoles, which is similar to the usual assumption that a charge does not interact with itself.

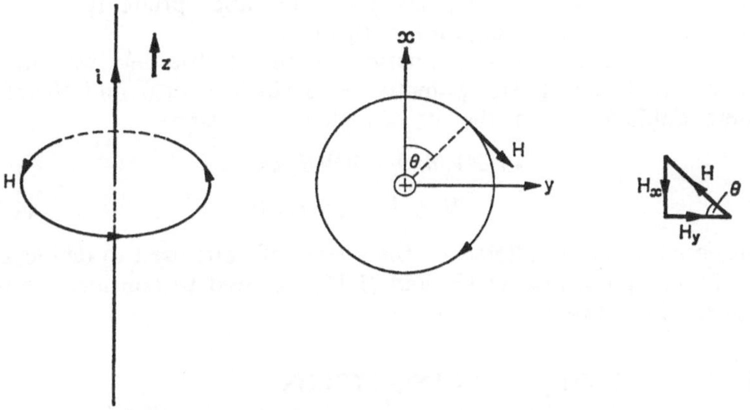

Figure 1-11. Field due to current in long straight wire.

We can find the field due to an infinitely long straight wire, either from first principles, by constructing an infinite magnetic shell and using Eq. (1.10), or by noting that the field must be cylindrically symmetrical so that the lines of force are circles centred on the wire. The change in potential on going around one of the lines of force will be:

$$H\times 2\pi r=4\pi i \text{ (from the previous argument)}$$

i.e.

$$H=2i/r \tag{1.12}$$

Now from the diagram

$$H_x=H\sin\theta=+2iy/r^2$$

and

$$H_y=H\cos\theta=-2ix/r^2$$

hence, differentiating, keeping r constant:

$$\partial H_y/\partial x - \partial H_x/\partial y = 4i/r^2 = 4\pi i^*$$

in vector language this becomes:

$$\nabla \wedge \bar{H} = \text{curl } \bar{H} = 4\pi \bar{\imath}^* \tag{1.13}$$

where $i^* = i/\pi r^2 =$ current density.

In S.I. units this relationship is

$$\text{curl } \bar{H} = \bar{\imath}^* \tag{1.13a}$$

This equation can be extended to the general case, either by considering systems of long straight wires appropriate for a given situation, or directly using Stokes' theorem.

For the corresponding equation for the electric field we can use the fact that the electric potential V is single valued and therefore differentiable (except at the site of a charge) i.e. since

$$\partial^2 V/\partial x \partial y = \partial^2 V/\partial y \partial x$$

$$\nabla \wedge \bar{E} = \text{curl } \bar{E} = 0 \tag{1.14}$$

This equation is not altered by the system of units used to define $\bar{E}$.

To complete Eqs. (1.13) and (1.14) we need to consider electromagnetic induction.

1.5. EFFECTS OF CHANGING FIELDS

From what we have seen so far it appears that electric fields arise from the existence of charges, and that magnetic fields arise from their motion. It follows from this that if we know the magnetic field due to a moving charge and the interaction of a (moving) charge with a magnetic field, then we can add these to the coulomb law of electrostatics and the law of conservation of charge to give us a complete picture of classical electromagnetism.

Magnetic field at P due to loop $BB' = i\, dS/r^3$ perpendicular to the page, according to Eqs. (1.10) and (1.15); i.e. field due to

$$BB' = i\, dy\, dr \sin\theta/r^3 = i\, dl\, dr \sin\theta/r_0 r^2$$

The field at P due to AA' is the sum of all such contributions, i.e. field at

$$P = \int i\, dl\, dr \sin\theta/r_0 r^2 = i\, dl \sin\theta/r_0^2$$

or in vector language,

$$\bar{H} = i\overline{dl} \wedge \bar{r}/r^3 \tag{1.15}$$

If the current is due to the movement of a single point charge magnitude q e.s.u. $=q/c$ e.m.u. ($c=$ ratio of these two units); (in S.I. units there is only one value assigned to a charge and therefore $c=1$); then $i\,dl=q\bar{v}/c$, $\bar{v}$ being the velocity of the charge and we obtain:

$$\bar{H}=(q/c)\bar{v}\wedge\bar{r}/r^3 \tag{1.16}$$

Now the electric field at P due to the charge q is $q\bar{r}/r^3$, so we have the following relationship between the electric and magnetic fields:

$$\bar{H}=\bar{v}\wedge\bar{E}/c \tag{1.17}$$

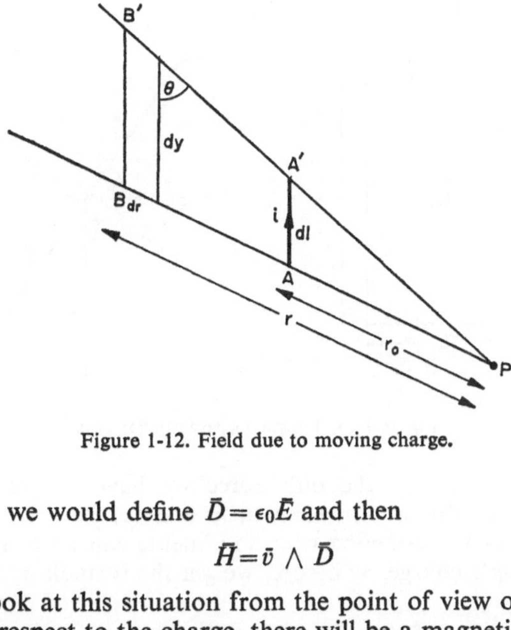

Figure 1-12. Field due to moving charge.

In S.I. units we would define $\bar{D}=\epsilon_0\bar{E}$ and then

$$\bar{H}=\bar{v}\wedge\bar{D} \tag{1.17a}$$

If we look at this situation from the point of view of an observer at rest with respect to the charge, there will be a magnetic force acting on a point magnet at P, which is now moving at $\bar{v}'=-\bar{v}$; i.e.

$$\text{magnetic force at moving point}=\bar{E}\wedge\bar{v}/c \tag{1.18}$$

In S.I. units the magnetic force at a moving point is:

$$(\mu_0\bar{H})=\mu_0\epsilon_0\bar{E}\wedge\bar{v}$$

Force on a Moving Charge in a Magnetic Field

What we have just found is in effect, the force which would be acting on a hypothetical magnetic pole moving in an electric field; now we find the corresponding expression for a charge moving in a magnetic field. Consider a rectangular coil in a magnetic field as in Figure 1-13.

$$\text{Torque about } z \text{ axis} = \mu B \sin\theta$$
$$= i\,ab\,B \sin\theta = f\,b \sin\theta$$

i.e.

$$\text{force on one side} = \bar{f} = i\bar{a} \wedge \bar{B} \qquad (1.19)$$

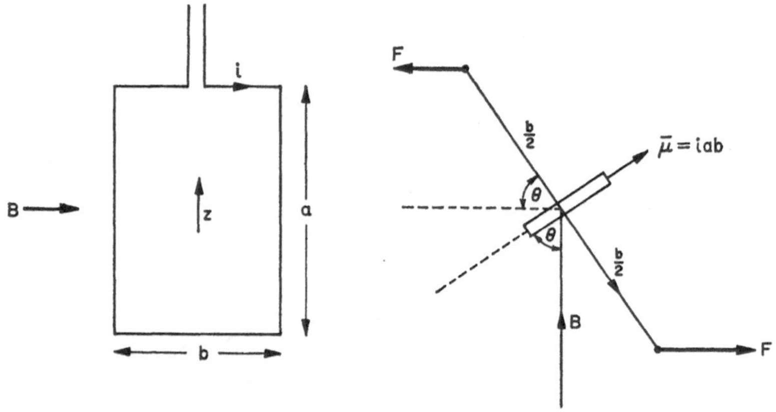

Figure 1-13. Torque on rectangular coil.

We have assumed that the only force we have to consider is that perpendicular to the current element *ia* and to the field. There could have been other forces acting whose moments cancelled. If the current is due to a single charge, as before, we get the formula for the Lorentz force:

$$\bar{f} = q\bar{v} \wedge \bar{B}/c \qquad (1.20)$$

It is worthwhile collecting the results in Eqs. (1.18) and (1.20) together, i.e. for a particle moving in steady electric ($\bar{E}$) and magnetic ($\bar{B}$) fields:

$$\left.\begin{aligned} \bar{F}_E &= \text{electric force} = \bar{E} + \bar{v} \wedge \bar{B}/c \\ \bar{F}_B &= \text{magnetic force} = \bar{B} - \bar{v} \wedge \bar{E}/c \end{aligned}\right\} \text{(per unit charge/pole)} \qquad (1.21)$$

In S.I. units these expressions become

$$\left.\begin{aligned} \bar{F}_E &= \bar{E} + \bar{v} \wedge \bar{B} \\ \bar{F}_B &= \bar{B} - (\bar{v} \wedge \bar{E})\epsilon_0\mu_0 \end{aligned}\right\} \qquad (1.21a)$$

It is worthwhile noting here that the velocity of light (i.e. electromagnetic waves) is,

$$c=\frac{1}{\sqrt{\epsilon_0\mu_0}}$$

Using Eqs. (1.2), (1.3), (1.13) and (1.14) we find:

$$\nabla \wedge \bar{F}_E=\text{curl } \bar{F}_E=-\bar{v}.\nabla\bar{B}/c=-d\bar{B}/dt\ (l/c)$$

$$\nabla \wedge \bar{F}_B=\text{curl } \bar{F}_B=+\bar{v}.\nabla\bar{E}/c=+d\bar{E}/dt\ (l/c)$$

where we have used the identity: $dG/dt=\partial G/\partial t+(\partial G/\partial x)\ dx/dt\ldots$

If now we change to a frame of reference which is stationary with respect to the point of measurement, $dG/dt=\partial G/\partial t$, $\bar{F}_E$ becomes $\bar{E}$, and curl $\bar{F}_B$ becomes curl $\bar{B}-4\pi\rho\bar{u}/c$; where $\bar{E}$ and $\bar{B}$ are now measured in the new frame. Now we have the last two of Maxwell's equations:

$$\left.\begin{array}{l} c\text{ curl } \bar{E}=-\partial\bar{B}/\partial t \\ c\text{ curl } \bar{B}=+\partial\bar{E}/\partial t+4\pi\rho\bar{u} \end{array}\right\}\ (u\ll c) \tag{1.22}$$

In terms of S.I. units these are:

$$\left.\begin{array}{l} \text{curl } \bar{E}=-\partial\bar{B}/\partial t \\ \text{curl } \bar{H}=+\partial\bar{D}/\partial t+\rho\bar{u} \end{array}\right\} \tag{1.22a}$$

This completes our introduction to electromagnetic theory. Many of the assumptions we have made are rather difficult to pin down and so often one starts from Maxwell's equations and using the special theory of relativity one can deduce the various formulae for electromagnetic interactions such as the Lorentz force or the inverse square laws as theorems. However, many of the formulae are still not in the most convenient form for converting the classical theory into quantum mechanics.

1.6. THE VECTOR POTENTIAL

When we come to discuss energies rather than forces it is convenient to introduce potential fields to replace $\bar{E}$ and $\bar{B}$. In the case of the electrostatic field this was easily done by making $\bar{E}$ the gradient of a scalar function $(-V)$. In the case of a magnetic field the situation is complicated by the fact that in general curl $\bar{B}\neq 0$, so that we cannot define $\bar{B}$ simply in terms of a scalar function.

We can, however, define the magnetic field in terms of a vector function $\bar{A}$ such that:

$$\bar{B}=\text{curl } \bar{A}=\nabla \wedge \bar{A} \tag{1.23}$$

Since div curl=0 this definition of $\bar{B}$ is essentially a restatement of equation (1.3), i.e. div $\bar{B}=0$.

Formulae often become simplified when written in terms of $\bar{A}$, for example the equation for curl $\bar{E}$ [see Eq. (1.22)], becomes:

$$\bar{E}=-\text{grad } V-(1/c)\, d\bar{A}/\partial t \tag{1.24}$$

Some useful functions to use for $\bar{A}$ are:

(i) for a steady, homogeneous field $\bar{B}$, $\bar{A}=\frac{1}{2}\bar{B}\wedge\bar{r}$ (1.25)

(ii) for a charge q moving at a steady velocity $\bar{v}$, the vector potential at distance r is

$$\bar{A}=(q/c)\bar{v}/r \quad \text{(c.g.s.)}$$

or

$$\bar{A}=q\bar{v}\mu_0/4\pi r \quad \text{(S.I.)} \tag{1.26}$$

(iii) at distance $\bar{r}$ from a dipole moment "$\bar{\mu}$" vector potential is given by

$$\bar{A}=\bar{\mu}\wedge\bar{r}/r^3 \quad \text{(c.g.s.)}$$

or

$$\bar{A}=\frac{\mu_0}{4\pi}\bar{\mu}\wedge\bar{r}/r^3 \quad \text{(S.I.)} \tag{1.27}$$

Generally it is some effect of the electromagnetic force field which we measure, so that the electrostatic potential V is found by integration. A rather more complicated integration of $\bar{B}$, will give us $\bar{A}$. In both cases there will be an arbitrary factor introduced by the integration and in the case of the scalar potential function V, this factor is fixed by defining a zero of energy. With the vector potential $\bar{A}$, the arbitrary factor can be the gradient of any scalar function since

$$\bar{B}=\text{curl } \bar{A}=\text{curl }(\bar{A}+\text{grad } X)$$

This follows from the identity curl grad=0.

The choice of X determines what we call the "gauge" of our mathematical formulation of a problem, and since the field of force is independent of the gauge, we have to make sure that the results we derive do not depend on a fortuitous choice of gauge. Very often it is convenient to work in what is called a Coulomb gauge, i.e. one in which

$$\text{div } \bar{A}=\nabla.\bar{A}=0 \tag{1.28}$$

1.7. THE MAGNETOGYRIC RATIO AND SPIN

A quantity which will be of great interest to us is the magnetogyric ratio, that is the effective magnetic moment of a circulating charge divided by its angular momentum.

We can easily calculate this ratio for a small circular current loop as in the diagram, i.e.

$$\text{magnetic moment} = i\,dS = i\pi r^2$$
$$= qv\pi r^2/2\pi rc = (q/2mc)mvr = (q/2mc)M$$

i.e.

$$\left.\begin{array}{ll}\text{magnetogyric ratio} = \gamma = q/2mc & \text{(c.g.s.)}\\ \gamma = q/2m & \text{(S.I.)}\end{array}\right\} \tag{1.29}$$

This formula is easily proved in the more general case when the angular momentum is a constant of the motion.

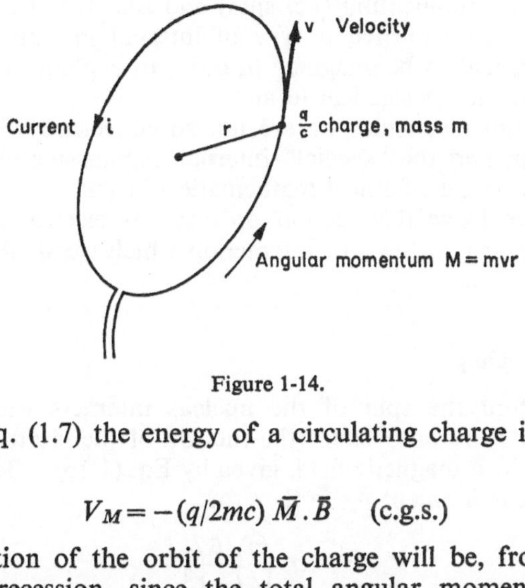

Figure 1-14.

From Eq. (1.7) the energy of a circulating charge in a magnetic field is:

$$V_M = -(q/2mc)\,\bar{M}.\bar{B} \quad \text{(c.g.s.)} \tag{1.30}$$

The motion of the orbit of the charge will be, from Eq. (1.6), a Larmor precession, since the total angular momentum remains constant with time when $\bar{\mu}$ is proportional to $\bar{M}$.

To determine the motion of a charge more exactly we cannot assume that its "orbit" is more or less fixed, really we should start from the equations for the force on a moving charge, i.e. (1.21):

$$\text{force} = m\dot{\bar{v}} = q\bar{E} + q\bar{v} \wedge \bar{B}/c$$

If now we write this in terms of a frame of reference rotating at an angular velocity of ω, it becomes:

$$m\dot{\bar{v}}_r + 2m\bar{\omega} \wedge \bar{v}_r + m\bar{\omega} \wedge (\bar{\omega} \wedge \bar{r}) = q\bar{E} + q\bar{v}_r \wedge \bar{B} + q(\bar{\omega} \wedge \bar{r}) \wedge \bar{B} \tag{1.31}$$

↖Coriolis force ↖centripetal force

where $\bar{v}_r$ is the velocity of the charge in the rotating frame. Usually we say that in weak fields the Coriolis term dominates, so that the centripetal force can be neglected. In this event the equation of motion in a frame rotating at $\omega = -qB/2mc$, is the same as that in the original (inertial) frame in the absence of a magnetic field; i.e. $m\dot{\bar{v}}_r = q\bar{E}$. This means that the orbit of the charge is precessing at the Larmor frequency, and is apparently the case for electronic orbital motion.

There is, however, another "extreme" possibility and that is when the Coriolis force is that which can be neglected. In this case the precession is at twice the Larmor frequency and the corresponding magnetogyric ratio is q/mc.

When it was found that fundamental particles could possess an intrinsic angular momentum (i.e. spin) and that for electrons $\gamma = q/mc$ a special, or rather particular type of internal motion, such as that just described, had to be imagined in order to explain the 'anomalous' magnetic moments in classical terms.

In quantum mechanics we are not so concerned with the details of the motion, and the "special" internal angular momentum can be introduced by means of a neat mathematical "trick".

Before we leave this section on classical electromagnetism it is useful to look at two types of interaction which we shall keep coming across later.

Spin–orbit Coupling

In an atom the spin of the nucleus interacts with the orbital motion of the electron because the motion of the electronic charge is accompanied by a magnetic field, given by Eq. (1.16). The interaction with the nuclear moment $\bar{\mu}_N$ is therefore

$$V_N = -\mu_N \left(\frac{q}{c}\right) \frac{\bar{v} \wedge \bar{r}}{r^3}$$

$$= +\mu_N \left(\frac{q}{mc}\right) \frac{m\bar{r} \wedge \bar{v}}{r^3}$$

But the charge on an electron is $-e$ and its mass "m", hence

$$V_N = -\frac{e}{mc} \frac{\bar{\mu}_N . \bar{L}}{r^3} \qquad \bar{L} = \text{orbital angular momentum} \tag{1.32}$$

From the point of view of the electron spin according to Eq. (1.22), it experiences a magnetic force field of

$-\bar{v} \wedge \bar{E}/c = ze(\bar{r} \wedge \bar{v}/cr^3)$ where ze = charge on the nucleus

Hence its potential energy should include a term

$$(\mu_e z e\bar{r} \wedge \bar{v}/cr^3) = +\frac{ze}{mc}\,\bar{\mu}_e.\bar{L}/r^3 \qquad (1.33)$$

Although Eq. (1.33) is of the correct form the expression is too large for the spin–orbit coupling energy by a factor of two—due to certain relativistic corrections. The correct expression is:

$$V_{LS} = -\frac{ze^2}{2m^2c^2}\frac{\bar{L}.\bar{S}}{r^3} \qquad (1.34)$$

We shall derive the corrected formula elsewhere from the Dirac relativistic equation for an electron.

Fermi Contact Interaction

Another term in the magnetic Hamiltonian which will concern us is the Fermi Contact Interaction which is derived from the interaction between a spherically symmetrical "cloud" of electron spin and the nuclear moment.

The easiest way to see how this arises is to start by representing the spins by magnetic dipoles.

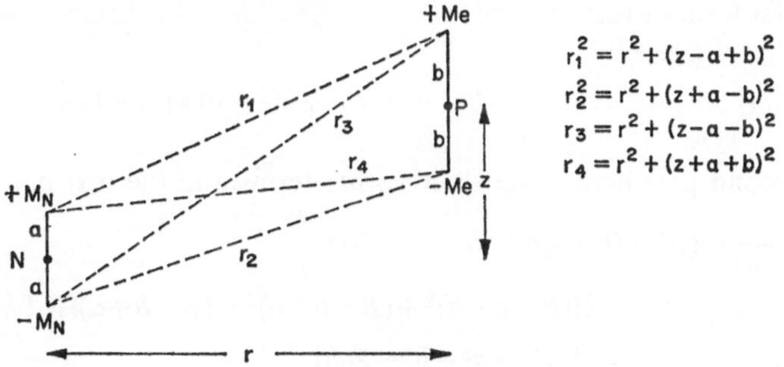

Figure 1-15.

At "*P*" the electron density is ρ.

So the moment of a small volume dV there, is

$$\mu_e\rho\, dV = 2bm_e\rho\, dV$$

The potential energy of the two dipoles is

$$dV = m_N m_e \rho\, dV\left[\frac{1}{r_1} + \frac{1}{r_2} - \frac{1}{r_3} - \frac{1}{r_4}\right] \qquad (1.35)$$

The natural system of coordinates to use are cylindrical polars when

$dV = r\,dr\,dz\,d\phi$ (ϕ = angle of rotation about z axis). Components of the dipole at P which are perpendicular to the nuclear moment give zero contribution to the potential energy on summation over the whole electron cloud and therefore we shall leave them out. The total potential energy of the electron cloud is the summation over all the electron cloud of the expression given in Eq. (1.35). This sum becomes an integral and integrating over ϕ first (from 0 to 2π) and using the relation $r\,dr = r_1 dr_1$ etc., we find:

$$V = 2\pi\, m_N m_e \iint \rho[dr_1 + dr_2 - dr_3 - dr_4]\, dz$$

We shall assume that ρ is constant over the range and if we integrate to a sphere radius "R" the results are particularly simple. The limits are:

$$0 \leqslant r \leqslant \sqrt{R^2 - z^2}$$

$$-R \leqslant z \leqslant +R$$

and the integral becomes:

$$2\pi m_N m_e \rho \int_{-R}^{+R} [r_1 + r_2 - r_3 - r_4]_{r=0}^{r=\sqrt{R^2-z^2}}\, dz$$

The first term of this integral is

$$\int_{-R}^{+R} [r_1]_{r=0}^{r=\sqrt{R^2-z^2}}\, dz = \int_{-R}^{+R} \{[R^2 + (b-a)^2 + 2(b-a)\,z]^{1/2} - (z + b - a)\}\, dz$$

The second part here cancels with other terms and the first part is:

$$\frac{1}{2(b-a)}\,\tfrac{2}{3}\Big[(R^2 + (b-a)^2 + 2(b-a)\,z)^{3/2}\Big]_{-R}^{+R}$$

$$= \tfrac{2}{3}[(R+b-a)^2 + (R-b+a)^2 + (R-b+a)(R+b-a)]$$

$$= \tfrac{2}{3}[3R^2 + a^2 + b^2 - 2ab]$$

The integral for r_2 is identical to this and those for r_3 and r_4 are both:

$$\tfrac{2}{3}[3R^2 + a^2 + b^2 + 2ab]$$

The overall integral is therefore independent of R and equal to:

$$\tfrac{4}{3}(-2ab - +2ab) = -\tfrac{4}{3}(2a)(2b)$$

The potential energy becomes, remembering that $2m_N a = \mu_N$, $2m_e b = \mu_e$:

$$-\frac{8\pi}{3}\mu_N \mu_e \rho(0) \qquad \text{(value of } \rho \text{ at } r=0,\ z=0 \text{ is } \rho(0))$$

The variation of ρ with r and z does not matter as long as it remains spherically symmetrical.

Chapter 2

Elements of Quantum Theory

2.1. HAMILTON'S FORM OF MECHANICS

When we are dealing with relatively simple systems, in which we can measure distances and time intervals accurately, the ideas of force, potential and kinetic energies, and so on, are meaningful and useful terms to employ in describing phenomena. In many cases, however, e.g. in atomic systems, we cannot analyse the situation in quite the same way, because it is rather difficult to locate the particles and specify their velocities at a given time. We are then thrown back on to the "conservation" laws in order to find "observables" or "measureables". The systems in which we shall be interested are those in which the total energy is conserved. When we write the total energy as the sum of the potential and kinetic energy functions, it is called the Hamiltonian of the system. It is then given the symbol H, so that just as we write y as a function of x in the form: $y=f(x)$, so we can write the total energy as a function of the various types of coordinates, e.g.

$$E=H(x, y, z, v_z, v_y, v_x, t) \tag{2.1}$$

The Newtonian and special relativistic functions for the total energy of a particle are:

(a) Newtonian

$$H=\tfrac{1}{2}mv^2+V(xyz) \qquad \text{(2.2)a}$$

(b) Special relativistic

$$H=m\beta c^2+V(xyz) \qquad \text{(2.2)b}$$

where m = rest mass and c = velocity of light. $\beta=(1-v^2/c^2)^{-1/2}$.

Rather than use velocities as the independent variables it is more convenient to use momenta as defined by Hamilton's canonical equations, i.e.

$$dp_r/dt = -\partial H/\partial q_r, \qquad dq_r/dt = +\partial H/\partial p_r \tag{2.3}$$

The q_r are space and time coordinates and the p_r the "conjugate" momenta. We can easily verify that for simple mechanical systems if $q_r = x$ then $p_r = mv_x$ and so on.

The first of Eqs. (2.3) expresses the force $d\bar{p}/dt$ in terms of the gradient of a function, so from the previous chapter we could expect difficulties when we deal with magnetism, since the magnetic field cannot usually be expressed in this way. These difficulties are, however, easily overcome. Now the force on a charge is

$$m d\bar{v}/dt = q\bar{E} + q\bar{v} \wedge \bar{B}/c$$
$$= [-\text{grad } V - 1/c . d\bar{A}/dt - \text{grad } (\bar{v} . \bar{A})]q$$

where we have used Eqs. (1.24) and (1.23) and are in a Coulomb gauge, i.e.

$$(d/dt)(m\bar{v} + q\bar{A}/c) = -q \text{ grad } (V - v . \bar{A}/c)$$

hence we can verify that $p_r = mv_r + qA_r/c$ and the equations for the energy become:

(a) Non-relativistic

$$H = (\bar{p} - q\bar{A}/c)^2/2m + V$$

(b) Relativistic

$$H = [(p - qA/c)^2 + m^2c^2]^{1/2}c + V \tag{2.4}$$

We have included the rest mass–energy in the Hamiltonian in the relativistic case.

2.2. TRANSITION TO QUANTUM MECHANICS

The laws of classical mechanics were, of course, deduced from experiments made on a certain scale, so it is not very surprising that they have to be modified when extrapolated to very large, or very small scale phenomena. In the case of molecular, atomic or subatomic systems, the mechanical laws have to include the fact that when we make a measurement on a system, our apparatus has to interact with it in some way; this interaction cannot be made negligibly small as we can generally assume in observations on macroscopic systems.

For microscopic systems, then, we need to use quantum mechanics, and the transition from classical theory can be made in a relatively simple way. First we have to abandon the idea of certainty and replace it by probability, thus the relative probability that a particle will be at a certain place (xyz) at a certain time is $|\psi(xyzt)|^2$, or in Dirac's notation $\langle|\delta(xyzt)|\rangle$, where for our purposes, ψ or $|\rangle$ is called the wavefunction of the particle. "δ" is Dirac's δ-function. Secondly, the conjugate momenta occurring in Eq. (2.3) are to be replaced by a corresponding differential operator, i.e.

$$p_r = (\hbar/i)\partial/\partial q_r \tag{2.5}$$

where $\hbar$ is Planck's constant divided by 2π, and $i=\sqrt{-1}$. A dynamical variable will be made up of some combination of these p_r's and q_r's, and will generally operate on an appropriate wavefunction.

Thirdly, the average or expectation value of a dynamical variable F, say, is to be given by:

$$\bar{F} = \int \psi^* F \psi d\tau \Big/ \int \psi^* \psi d\tau \tag{2.6}$$

where ψ^* is the complex conjugate of ψ and $d\tau$ an element of volume in the space of $6N$ dimensions, if N is the number of particles. In Dirac's notation this equation is:

$$\bar{F} = \langle|F|\rangle / \langle|\rangle = \langle F\rangle / \langle\ \rangle \tag{2.7}$$

We can see from these definitions that Dirac's formulation is in terms of integrals and that the δ-function has to be introduced as a means of differentiating to give a quantity such as $\psi^*\psi$, the probability per unit volume at a particular place.

Now we can set up Schrodinger's wave equation as an example of how to formulate a problem in quantum mechanics. In classical terms, $E = H = p^2/2m + V(xyz)$. In quantum mechanical terms,

$$E|\rangle = H|\rangle = p^2/2m|\rangle + V|\rangle$$

This becomes, on using Eq. (2.5) in conjunction with Eq. (2.3):

$$H|\rangle = -(\hbar^2/2m)\nabla^2|\rangle + V|\rangle = E|\rangle = i\hbar\,(d/dt)|\rangle \tag{2.8}$$

The method in quantum mechanics is to find which values of an operator, such as the energy in Eq. (2.8), are compatible with the boundary conditions of the problem; these values are called the eigenvalues of the operator and the corresponding wavefunctions are its eigenfunctions.

It is the Eigenvalues of an Operator, which we Observe

If a system is in an eigenstate of a dynamical variable A and also of B then we use as labels the appropriate eigenvalues a and b, say, in the wavefunctions, i.e.

$$A \mid a, b\rangle = a \mid a, b\rangle$$
$$B \mid a, b\rangle = b \mid a, b\rangle$$

2.3. COMMUTATION RELATIONS

It is very useful to label the possible states of a system with the values of the various variables which describe it. In classical mechanics this could be done using the space and velocity/momentum coordinates, six for each particle. In quantum theory this can no longer be done and one way of expressing this is that "they do not commute".

If two variables can be observed at the same time then the system must be in a state which is an eigenstate with respect to both of them, since we can only observe their eigenvalues. Assuming that numbers commute with all operators and, of course, with each other, then in the example of the previous section,

$$AB \mid a, b\rangle = Ab \mid a, b\rangle = bA \mid a, b\rangle$$
$$= ba \mid a, b\rangle = ab \mid a, b\rangle = \ldots BA \mid a, b\rangle$$

i.e.

$$(AB - BA) \mid a, b\rangle = 0$$

or $AB - BA = 0$, and A and B commute. The inverse is also true, i.e. if the operators do not commute then they cannot be observed simultaneously.

We can quickly see that the space and momentum operators do not commute, for:

$$p_r q_r - q_r p_r = (\hbar/i)[(d/dq_r)q_r - q_r(d/dq_r)]$$
$$= (\hbar/i) \neq 0 \tag{2.9}$$

This can be written in a more condensed form, i.e.

$$(p_r q_s - q_s p_r)/(\hbar/i) = [\bar{p}, \bar{q}] = \delta_{rs} \tag{2.10}$$

where δ_{rs} is zero for $r \neq s$, and unity for $r = s$. The expression $[p, q]$ is the "quantum Poisson brackets" and is analogous to the Poisson brackets of classical theory.

Angular Momentum

The components of the angular momentum of a system do not commute with each other, and are given by the same equation as in classical theory, i.e.

$$\bar{M}=\bar{r}\wedge\bar{p}$$

i.e.

$$\begin{aligned}M_x&=yp_z-zp_y\\M_y&=zp_x-xp_z\\M_z&=xp_y-yp_x\end{aligned}\tag{2.11}$$

(see Figure 2.1).

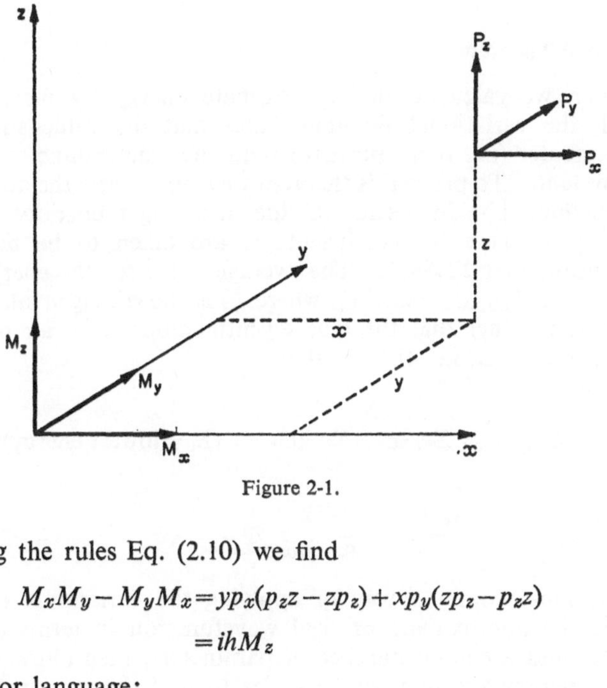

Figure 2-1.

Using the rules Eq. (2.10) we find

$$\begin{aligned}M_xM_y-M_yM_x&=yp_x(p_zz-zp_z)+xp_y(zp_z-p_zz)\\&=i\hbar M_z\end{aligned}$$

or in vector language:

$$\bar{M}\wedge\bar{M}=i\hbar\bar{M}\tag{2.12}$$

2.4. APPROXIMATE METHODS

"Idealistically" we have outlined in the previous sections how to set up and solve mathematically, any dynamical problem we care to define. The method is to write down the classical equations of motion, replace the variables by corresponding operators and then to solve the resulting differential equations for their eigenvalues and eigenfunctions.

In practice, it is not generally possible to carry out all of these operations, since the interesting problems are seldom sufficiently simple. Usually we have to try and deduce the properties of real systems by comparing them with idealized ones, in one way or another. This comparison is achieved mathematically by writing the wavefunction of the real system as a linear combination of those of the idealized ones. An approximate value of any variable can then be calculated by means of Eqs. (2.6) or (2.7).

Variational Theorem

When we calculate the approximate energy by means of this method, the variational theorem states that the value so obtained will be greater (i.e. more positive) than the true ground-state energy of the system. To prove this theorem we simply write the approximate wavefunction, $|\rangle$, in terms of the true eigenfunctions $|r\rangle$; i.e. $|\rangle = \Sigma c_r |r\rangle$ where the coefficients c_r are taken to be normalized, which means that $\Sigma c_r^2 = 1$. The average value for the energy is then $\langle | H | \rangle$, i.e. $E_{\text{approx.}} = \Sigma c_r^2 E_r$, where E_r is the rth eigenvalue, and we have used the fact that the true eigenfunctions form an orthogonal (and complete) set, i.e. $\langle r | s\rangle = 0 \; r \neq s$.

Hence

$$E_{\text{approx.}} = \Sigma c_r^2 E_r \geqslant \Sigma c_r^2 E_0 \qquad (E_0 = \text{lowest energy})$$

i.e.

$$E_{\text{approx.}} \geqslant E_0.$$

The most common way of applying the variational theorem is to write the approximate or trial wavefunction in terms of the coordinates and a chosen number of parameters, then allow the values of these parameters to vary, in order to find the least value of the energy integral, Eq. (2.6), i.e. if the trial function is $| \alpha_1, \alpha_2, \alpha_r \ldots \rangle$, then we apply the condition $(\partial/\partial\alpha_r)\langle \ldots \alpha_r \ldots | H | \ldots \alpha_r \ldots \rangle = 0$, for each parameter "$\alpha_r$". In the more familiar examples, the parameters are simply the coefficients in a linear series of appropriate "basis" functions, i.e. approximate wavefunction $= \Sigma c_r \times r$th basis function. We choose these basis functions in such a way as to minimize the number of terms in the series and make the problem tractable, yet still providing a reasonable description of the system considered.

Perturbation Theory

When the basis functions are the eigenfunctions of a simplified Hamiltonian H_0 which is almost the same as the Hamiltonian H_0+H' of the system of interest, then we can use an approximate form of the variational method called perturbation theory. Thus if

$$H_0 \mid r\rangle = E_r \mid r\rangle,$$

then our first trial wavefunction will be $\mid 0\rangle$ and the energy of the system is calculated to be:

$$E=\langle 0 \mid H_0+H' \mid 0\rangle = E_0+\langle 0 \mid H' \mid 0\rangle \qquad (2.13)$$

where H' is called the perturbation and last term, the integral of H', the first-order perturbation energy.

Often Eq. (2.13) is not very good for calculating specific effects, so we then have to improve our trial function, i.e. by taking a linear combination of the $\mid r\rangle$. The ground-state energy is now exactly given if we use an infinite number of terms,

$$(H_0+H'-E)\Sigma c_r \mid r\rangle = 0$$

We are interested in the ground-state energy which is approximately E_0, so this equation gives:

$$\Sigma c_r(E_r-E_0+H') \mid r\rangle \approx 0$$

if H' is small, then all of the c_r will be small except c_0, hence,

$$c_r/c_0 \approx \langle 0 \mid H' \mid r\rangle/(E_0-E_r), \qquad r\neq 0 \qquad (2.14)$$

Now we can get the expression for the second-order perturbation energy using as our wavefunction $\Sigma c_r \mid r\rangle$, i.e. form Eq. (2.7):

$$E^2_{\text{pert.}} = \Sigma_r\langle 0 \mid H' \mid r\rangle\langle r \mid H' \mid 0\rangle/(E_0-E_r), \qquad \left(\begin{matrix} r\neq 0 \\ c_0\approx 1\end{matrix}\right) \qquad (2.15)$$

2.5. THE INTERACTION BETWEEN RADIATION AND MATTER

In order to deal with this subject properly, we should consider the system of constant energy, "radiation+matter". This would involve us in some theory of quantum electrodynamics. For our purposes, however, it will be sufficient to treat the problem as an example in time-dependent perturbation theory.

Let us consider the transition between two states, labelled r and s, which are stationary states in the absence of an applied field, represented by $H'(t)$, i.e.

$$H_0 \mid r\rangle = E_r \mid r\rangle$$

$$H_0 \mid s\rangle = E_s \mid s\rangle$$

where $| r\rangle$ and $| s\rangle$ are independent of the time. The total wavefunction for a stationary state can be split up into two parts, i.e.

$$| r, t\rangle = a_r(t) . | r\rangle$$

then from Eq. (2.8) we find

$$a_r(t) = \text{constant} \times \exp(-E_r t / i\hbar) \tag{2.16}$$

In the presence of the radiation field the states of the system are no longer strictly stationary, effectively the original states get "mixed" and we can approximate to the eigenfunctions for the new situation by taking linear combinations of the "old" stationary states. For simplicity we shall only consider the contribution arising from direct transitions, i.e. those not involving states other than $| r\rangle$ and $| s\rangle$, thus we write for the new, time dependent wavefunctions:

$$| R\rangle = c_r \exp(-iE_r t/\hbar) \, | r\rangle + c_s \exp(-iE_s t/\hbar) \, | s\rangle \tag{2.17}$$

where for $| R\rangle$ we assume that in the presence of the small perturbation $H'(t)$, $c_r \gg c_s$; there is a similar equation for $| S\rangle$, but with $c_s \gg c_r$. For a system originally in state $| r\rangle$, we can write the relation:

$$\{H_0 + H'(t)\} \, | r\rangle \approx i\hbar(\partial/\partial t) \, | r\rangle$$

which leads to the following expression when we neglect second-order terms:

$$\partial c_s/\partial t = \exp[(E_r - E_s)t/i\hbar] \langle s \mid H' \mid r\rangle / i\hbar \tag{2.18}$$

We can integrate this last equation once we have decided on the form of $H'(t)$. If the field varies sinusoidally we can represent it by $H' = F'(\bar{r}) \exp(2\pi i \nu t)$, where ν is its frequency and F' an electric or magnetic vector. Now integrating Eq. (2.18) we obtain:

$$c_s = \langle s \mid F' \mid r\rangle \exp[(E_r - E_s - 2\pi\nu\hbar)t/i\hbar]/(E_r - E_s - 2\pi\nu\hbar) \tag{2.19}$$

Hence c_s will only be large when $E_r - E_s \approx h\nu$ — the Bohr frequency condition. The other term which determines c_s is $F_{sr}' = \langle s \mid F' \mid r\rangle$, and this is what gives us the selection rules for a given transition. Since c_s^2 is the probability that the system will be in state $| s\rangle$ at time t, it will represent the probability of a transition when the initial state is represented by $| r\rangle$.

2.6. THEORIES OF MOLECULAR STRUCTURE

The Magnitudes of effects which are observed in magnetic resonance spectra depend on the states and possible states of the systems under investigation. For example, the coupling of two spins depends on how their magnetic effects are transmitted through the molecule in which they are situated as well as their direct dipole–dipole interaction.

In order to avoid complication, the ground-state, and for that matter, the excited states, of a molecule are found neglecting magnetic interactions, which are comparatively small and can be estimated afterwards using perturbation theory. The first step, then, is to use one of the theories of molecular structure and for us it is sufficient to employ the most general approaches which are reasonably easy to apply to many molecules.

First there is the familiar orbital approximation in which each electron is assigned a wavefunction on the basis that we average out the repulsive effects of the other electrons. Thus each electron is thought of as if it were moving in the field of the nucleus and a cloud of negative charge. According to the Pauli principle the total wavefunction must be antisymmetric with respect to exchange of any two electrons and this leads to the idea of assigning at most two electrons, of opposite spins, to each spatial orbital.

L.C.A.O. Approach

If the possible orbitals of an electron in a molecule are written in terms of those of the constituent atoms, then we are assured that its behaviour when near a particular nucleus is almost the same as in the corresponding atom. Thus a molecular orbital may be a Linear Combination of Atomic Orbitals.

$$\psi_{MO} = \Sigma c_r \phi_r \qquad \phi_r \text{ is the } r\text{th atomic orbital (AO)}$$

If "H" is the one electron Hamiltonian for the molecule the molecular orbitals (MO's) of the molecule are found by minimizing the integral

$$E_{MO} = \int \psi_{MO} H \psi_{MO} dr, \text{ using the } c_r \text{ as parameters}$$

i.e.

$$\frac{\partial E}{\partial c_r} = 0$$

This leads to the secular equations:

$$\sum_s c_s(\beta_{rs} - ES_{rs}) = 0 \qquad r = 1, 2 \ldots \qquad (2.20)$$

Where

$$\beta_{rs} = \int \phi_r H \phi_s \, dr \text{ (resonance integral)} \qquad r \neq s$$

$$\beta_{rr} = \alpha_r \text{ (coulomb integral)}$$

$$S_{rs} = \propto \phi_r \phi_s \, dr = \text{overlap integral}$$

$$S_{rr} = 1$$

α_r is approximately the energy of atomic orbital ϕ_r.

The secular Eqs. (2.20) are solved by eliminating the coefficients when we get for non-trivial cases:

$$\det \mid \beta_{rs} - ES_{rs} \mid = 0 \tag{2.21}$$

From this equation we get the energy in terms of the integrals, and back substitution into Eq. (2.20) yields the coefficients. The ground-state of the molecule is then found by assigning electrons two at a time to the MO's starting from the lowest (Aufbau principle), as in the atomic orbital description of atoms. The assignment of electrons to MO's gives a "configuration" whose wavefunction is a determinant to allow for the antisymmetrical properties of electrons.

i.e. If the occupied MO's are $\psi_0\psi_1 \ldots \psi_{N-1}$ the MO wavefunction Ψ for the molecule and there are $2N$ electrons.

$$\Psi = \frac{1}{\sqrt{2N!}} \begin{vmatrix} \psi_0\alpha(1)\psi_0\beta(1)\psi_1\alpha(1)\psi_1\beta(1) \ldots \\ \psi_0\alpha(2)\psi_0\beta(2) \ldots \end{vmatrix} \tag{2.22}$$

Where $\psi_0\alpha(1)$ mean that electron 1 is in MO ψ_0 with spin α, and so on.

Valence Bond Theory

In contrast to orbital theories, the theory of spin states start at the outset by considering all of the electrons at once. Each valence shell atomic orbital is assigned a spin wavefunction and one electron (if we do not consider "ionic" terms) so that the wavefunction of each spin state is a determinant, which ensures antisymmetry with respect to electron exchange. For example if there are four electrons in four atomic orbitals ϕ_a, ϕ_b, ϕ_c, ϕ_d the following spin rates are possible when the total spin is zero.

$$\alpha\alpha\beta\beta \qquad \beta\beta\alpha\alpha \qquad \alpha\beta\beta\alpha \qquad \alpha\beta\alpha\beta \qquad \beta\alpha\beta\alpha \qquad \beta\alpha\alpha\beta$$

Where $\alpha\alpha\beta\beta$ stands for N^{-1} det $(\phi_a\alpha, \phi_b\alpha, \phi_c\beta, \phi_d\beta)$ etc. N is a normalization factor.

Each one of these determinantal wavefunctions is a possible trial wavefunction for the system and we can get a better approximation by taking a linear combination of all six of them using the coefficients as parameters and invoking the variational theorem.

The number of spin-states increases rapidly with the number of valence shell atomic orbitals so a limitation is imposed on the trial wavefunctions. The spin states are grouped into sets which correspond to electron pair bonds between pairs of atomic orbitals. The valence-bond wavefunction is then the best linear combination of these sets; e.g. in the four orbital case there are only two independent sets I and II say (III is a linear combination of these) represented as bonds in what are called "canonical structures".

I II III

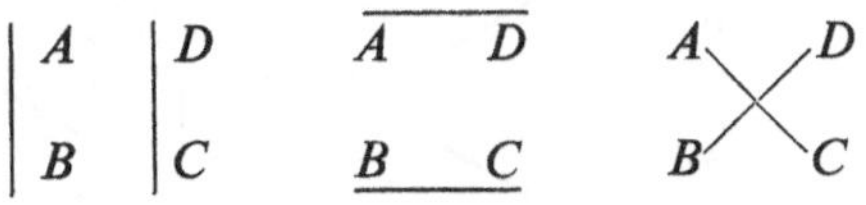

spin states

I		II		III	
α	α	α	β	α	α
β	β	α	β	β	β
α	β	α	β	α	β
β	α	β	α	α	β
β	α	β	α	β	α
α	β	α	β	β	α
β	β	β	α	β	β
α	α	β	α	α	α

$$\begin{aligned}\psi_{\rm I}&=N^{-1}\,(\alpha\beta\alpha\beta+\beta\alpha\beta\alpha-\alpha\beta\beta\alpha-\beta\alpha\alpha\beta)\\ \psi_{\rm II}&=N^{-1}\,(\alpha\beta\alpha\beta+\beta\alpha\beta\alpha-\alpha\alpha\beta\beta-\beta\beta\alpha\alpha)\\ \psi_{\rm III}&=N^{-1}\,(\alpha\beta\beta\alpha+\beta\alpha\alpha\beta-\alpha\alpha\beta\beta-\beta\beta\alpha\alpha)=\psi_{\rm II}-\psi_{\rm I}\end{aligned} \tag{2.23}$$

The valence bond wavefunction is:

$$\psi_{VB}=c_1\psi_{\rm I}+c_2\psi_{\rm II}$$

The secular equations are obtained from this in the usual way and in this example are:

$$c_1\int\psi_{\rm I}(H-E)_{\rm I}\,\psi dt+c_2\int\psi_{\rm I}(H-E)\,\psi_{\rm II}dt=0$$

$$c_1\int\psi_{\rm II}(H-E)\,\psi_{\rm I}dt+c_2\int\psi_{\rm II}(H-E)\,\psi_{\rm II}dt=0$$

In this simple example it is lengthy but not difficult to work out the matrix elements involved. In large systems we can adopt the following procedure:

(1) Draw a circle and mark on its circumference as many places as there are atomic orbitals.

(2) Join pairs of these points up so that none of the lines cross. When this has been done in all possible ways the diagrams (Rumer diagrams) represent a complete set of canonical structures.

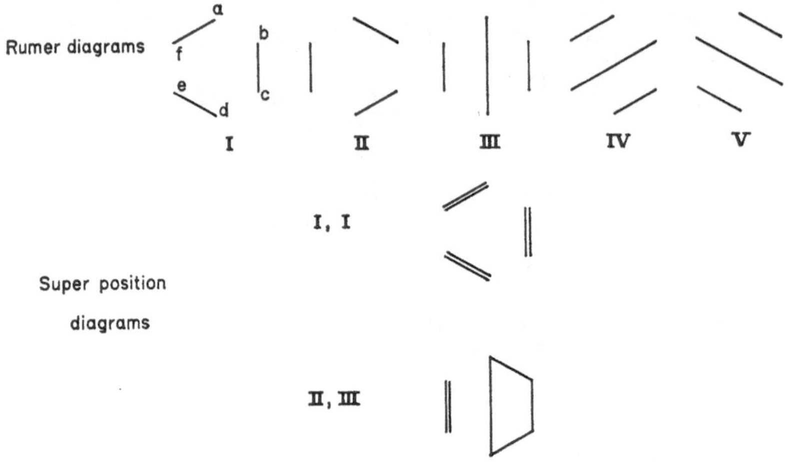

Figure 2-2. Rumer diagrams for 6 atomic orbital system.

(3) To find the matrix elements superimpose corresponding diagrams; e.g. for H_{II} $(=\int\psi_{\text{I}}(H-E)\psi_{\text{I}}dt)$ see Figure 2-2.

(4) Each closed figure on the superposition diagrams is called an island. If "n" is the number of atomic orbitals and "r" the number of islands, the matrix elements are given by the formula

$$\tfrac{1}{2}^{(n-r)}\Bigg[Q-E+\underset{\substack{r<s\\ r,\,s\text{ orbitals on same island separated by \textit{odd} number of bonds}}}{\sum J_{rs}}\;-2\underset{\substack{r<s\\ r,\,s\text{ orbitals on same island separated by \textit{even} number of bonds}}}{\sum J_{rs}}\;-\tfrac{1}{2}\underset{\substack{r<s\\ r,\,s\text{ on different islands}}}{\sum J_{rs}}\Bigg] \tag{2.24}$$

Where $Q = \int \phi_a(1),\ \phi_b(2),\ \phi_c(3) \ldots H\phi_a(1),\ \phi_b(2),\ \phi_c(3) \ldots dt$
(Coulomb integral)

$J_{ab} = \int \phi_a(1),\ \phi_b(2),\ \phi_c(3) \ldots H\phi_a(2),\ \phi_b(1),\ \phi_c(3) \ldots dt$
(exchange integral)

$S_{rs} = 0$ (neglect overlap)

In our 6 AO example $n=3$, for H_{II} $r=3$, for $H_{\mathrm{I\ III}}$ $r=2$ and so on. So for H_{II} the matrix element is

$$Q - E + J_{bc} + J_{de} + J_{fa} - \tfrac{1}{2}(J_{ab} + J_{cd} + J_{ef}) - \tfrac{1}{2}(J_{ac} + J_{ad} + J_{ae}) - \tfrac{1}{2}(J_{bd} + J_{be} + J_{bf}) - \tfrac{1}{2}(J_{ce} + J_{cf}) - \tfrac{1}{2}J_{df}$$

When there is an odd number of orbitals a "phantom" orbital is introduced and the problem worked out with the interactions between this orbital and the others being zero. Of course the odd electron system will have a resultant spin, i.e. opposite to that of the phantom orbital.

2.7. QUANTUM THEORY OF SPIN

In order to explain the splitting of lines in atomic spectra, both in and out of a magnetic field, and the deviation of atomic beams, when passing through inhomogeneous magnetic fields, we have to assume that electrons and nuclei can have intrinsic magnetic moments and presumably, corresponding angular momenta. (We have discussed this already in Chapter 1, section 1.7.)

It is not very surprising that we have the same difficulties in including this "spin" in quantum mechanics as in a classical framework, since the only difference between the two approaches is essentially in the commutation relations such as Eq. (2.10).

Whatever method we use, it seems we have to make two "extra" assumptions when we try to incorporate spin into our theory; one is that a particle has some sort of internal motion, more or less independent of its motion in space which we can observe, and the other is that this motion (internal) gives it a particular magnetic moment. However we imagine such a motion (e.g. as movement of the particle in its own "private" space), we have to represent it by a further set of independent variables, and these variables have to be introduced in such a way as to produce the correct magnetogyric ratio. In quantum theory we must also get the correct eigenvalues for the spin angular momentum.

A particularly convenient way of introducing spin into quantum mechanics is made possible because momenta are represented by differential operators. First we ascribe to the extra degrees of freedom an operator $\bar{\alpha}$ which will have three components and will commute with the spacial coordinates and their conjugate momenta. Secondly we introduce it into the Hamiltonian in the following way:

$$H=(\bar{\alpha}.m\bar{v})^2/2m+V(xyz) \tag{2.25}$$

Feynmann calls this the Pauli equation and in the absence of a magnetic field, no effects of the spin are observed so in that case it has to reduce to (2.2a). The operator $\bar{\alpha}$ therefore has to satisfy the conditions:

$$\alpha_x{}^2=\alpha_y{}^2=\alpha_z{}^2=1$$

$$\alpha_x\alpha_y+\alpha_y\alpha_x=0 \text{ etc} \tag{2.26}$$

i.e.

$$\alpha_r\alpha_s+\alpha_s\alpha_r=2\delta_{rs} \qquad \begin{pmatrix}\delta_{rs}=0 & r\neq s\\ \delta_{rs}=1 & r=s\end{pmatrix}$$

Relativistic Theory

To obtain the exact equivalent of Eq. (2.25) in relativistic terms, we would simply replace $m\beta\bar{v}$ by $\bar{\alpha}.m\beta\bar{v}$ in Eq. (2.2)b or in Eq. (2.4); however, Dirac found that the relativistic Hamiltonian for a particle with spin could be written more conveniently in a linear form, i.e.:

$$H=c(\bar{\alpha}.m\beta\bar{v}+mc\alpha_4)+V(xyz) \tag{2.27}$$

In the absence of fields this equation has to reduce to Eq. (2.4) and we find that the components of $\bar{\alpha}$ have to satisfy exactly the same conditions Eq. (2.26) as in the non-relativistic case.

We shall return to the Dirac equation, [Eq. (2.27)] later, for the moment we shall consider the Pauli equation [Eq. (2.25)].

In the presence of a magnetic field the components of the velocity no longer commute. Leaving out the potential energy V we can write Eq. (2.25) as follows:

$$H=\tfrac{1}{2}mv^2+\tfrac{1}{2}m\alpha_x\alpha_y(v_xv_y-v_yv_x)+\alpha_y\alpha_z(v_yv_z-v_zv_y)+\alpha_z\alpha_x(v_zv_x-v_xv_z)$$

where we have made use of the relations Eq. (2.26). Now substituting for $\bar{v}$ we can use the results of Eq. (2.1), i.e.

$$v^2=(1/m^2)(p-q/cA)^2=p^2/m^2-(\bar{p}.\bar{A}+\bar{A}.\bar{p})q/\text{cm}^2+q^2A^2/m^2c^2$$

$$\bar{v}\wedge\bar{v}=(\bar{p}\wedge\bar{A}+\bar{A}\wedge\bar{p})/m^2=(\bar{p}\wedge\bar{A})/m^2=(\text{curl }\bar{A})\hbar/im^2=\bar{B}\hbar/im^2$$

and we get for the energy in a magnetic field:

$$H=\tfrac{1}{2}mv^2-(q/mc)(\hbar \mid 4i)\bar{\alpha}\wedge\bar{\alpha}.\bar{B} \tag{2.28}$$

Now we put $\bar{S}=\hbar\bar{\alpha}\wedge\bar{\alpha}/4i$, (i.e. $S_z=\hbar\alpha_x\alpha_y/2i$, etc.), so that for a steady field $\bar{B}$, we obtain:

$$H=p^2/2m-(q/2mc)(\bar{L}+2\bar{S}).\bar{B}+q^2\bar{A}.\bar{A}/2mc^2 \tag{2.29}$$

where $\bar{L}$ is the orbital angular momentum.

We can call $\bar{S}$ the spin angular momentum because its components satisfy the same commutation relations as those of the orbital angular momentum [see Eq. (2.12.)], i.e.

$$\bar{S}\wedge\bar{S}=i\hbar\bar{S} \tag{2.30}$$

Both $\bar{L}$ and $\bar{S}$ commute with the non-relativistic Hamiltonian in Eq. (2.25) but not that in Eq. (2.27). The total angular momentum $\bar{J}=\bar{L}+\bar{S}$ commutes with both Hamiltonians in the absence of an electric field.

The value of S_z is easily found from its definition, i.e.

$$S_z{}^2=-\tfrac{1}{4}\hbar^2\alpha_x\alpha_y\alpha_x\alpha_y/i^2=-\tfrac{1}{4}\hbar^2\alpha_x{}^2\alpha_y{}^2/(-1)$$

i.e.

$$S_z=\pm\tfrac{1}{2}\hbar \tag{2.31}$$

Reduction of the Dirac Equation

There are several disadvantages in using the Pauli equation, one being that it is strictly applicable at velocities which are low compared with that of light, another being that it has to be reformulated in order to get the spin–orbit coupling energy. The Dirac equation does not suffer from these difficulties.

We can derive both the spin–orbit and Fermi contact interactions by deriving the non-relativistic Hamiltonian from Eq. (2.27) by the following steps

$$(H-V)(H-V)=(H-V)(E+mc^2-V)=(E+mc^2-V)^2-c\bar{\alpha}.\bar{p}V$$

we have used the fact that $\bar{p}$ is a differential operator, hence

$$(E+mc^2-V)^2=m^2c^2(\bar{\alpha}.\beta\bar{v}+\alpha_4c)^2+c\bar{\alpha}.\bar{p}V \tag{2.32}$$

if $2mc^2\gg|E-V|$, we find, leaving out the relativistic corrections

$$E-V=\tfrac{1}{2}mv^2-(q/mc)\bar{S}.\bar{H}+\bar{\alpha}.\bar{p}V/2mc \tag{2.33}$$

This equation is the same as that corresponding to Eq. (2.29) except for the last term which contains the two interactions we require.

Now the operator $\bar{\alpha}$ is best represented by an appropriate matrix, which means that Eqs. (2.25) and (2.27) are matrix equations so that they each represent a number of wave equations containing more than a single eigenfunction. In order to find one of the eigenfunctions we have to combine the set of equations so as to eliminate the others (e.g. see Slater). Rather than do this, we shall take a short cut to get an expression for the spin–orbit coupling energy.

First we take for example

$$\alpha_x = \tfrac{1}{2}(\alpha_x\alpha_y{}^2 + \alpha_x\alpha_z{}^2)$$

$$= \tfrac{1}{2}(\alpha_x\alpha_y\alpha_y - \alpha_z\alpha_x\alpha_z)$$

i.e.

$$(\hbar/i)\alpha_x = S_z\alpha_y - S_y\alpha_z = -\alpha_y S_z + \alpha_z S_y$$

i.e.

$$(\hbar/i)\alpha_x = -(\bar{S} \wedge \bar{\alpha})_x$$

similarly for the components.

Secondly, consider

$$Hx - xH = i\hbar\ \partial x/\partial t = i\hbar v_x$$

$$= c(\alpha.\bar{p}x - x\alpha.\bar{p})$$

$$= c\alpha_x(\hbar/i)$$

i.e.

$$-\bar{\alpha} = (\bar{v}/c)I$$

where I is an appropriate unit operator.

Using these two results the last term in Eq. (2.33) becomes:

$$\alpha.pV/2mc = -\bar{S} \wedge \bar{\alpha}.\bar{E}'/2mc \qquad \text{where } \bar{E}' = -\nabla V$$

$$= +(q/2mc^2)\bar{S}.\bar{v} \wedge \bar{E} \tag{2.34}$$

where the electric field strength $\bar{E} = \bar{E}'/q$, and $\bar{a}.\bar{b} \wedge \bar{c} = \bar{a} \wedge \bar{b}.\bar{c}$.

This formula [see Eq. (2.00)] is that for the spin–orbit coupling and for an electron, charge $-e$, in a central field given by $\bar{E} = \bar{r}f(r)$ it becomes:

$$H_{SL} = +(e/2m^2c^2)f(r)[\bar{L}.\bar{S} + (e/c)\bar{S}.\bar{r} \wedge \bar{A}] \tag{2.35}$$

It is the last part of this expression which leads to the Fermi contact interaction, in which case we can no longer neglect the potential energy V compared with $2mc^2$ when the electron is very close to the nucleus.

Chapter 3

The Observation of Magnetic Resonance

3.1. THE RESONANCE CONDITION

A particle with spin $\frac{1}{2}$ will exist in two energy states in a given magnetic field, each associated with a total spin angular momentum of magnitude $(\sqrt{3}/2)\hbar$. The component of the spin along the direction of the field will be $\pm\frac{1}{2}\hbar$, and the corresponding energy, $\mp\frac{1}{2}\hbar\gamma B_z$, taking the direction of the field as the z-direction.

The angular velocity associated with the electromagnetic field which most easily induces transitions between these levels is given in section 2.5, i.e.

$$\Delta E = h\nu = \hbar\omega = \hbar\gamma B_z$$

i.e.

$$2\pi\nu = \omega = \gamma B_z \tag{3.1}$$

This is the same angular velocity as that associated with the Larmor precession of the spin, as we can verify from Eq. (1.6), (also see section 1.7), and this is why the term "resonance" is given to the absorption/emission of radiation due to transitions between magnetic energy levels.

The same formula applies to assemblies of particles with spin $> \frac{1}{2}$ because differences between the z-components of the total angular momentum are always integral numbers of $\hbar$.

Since we know the various constants occurring in Eq. (3.1) from atomic spectra, we can work out the field-frequency relationship for, say, protons and electrons:

For a field of 10,000 gauss,

$$\left.\begin{aligned} \nu_{\text{proton}} &= 42.6 \text{ MHz} \\ \nu_{\text{electron}} &= 28{,}026 \text{ MHz} \end{aligned}\right\}$$

Generally proton resonances are observed at frequencies of 40–100 Mc/s and electron resonances at about 10,000 Mc/s (X-band) or 35,000 Mc/s (Q-band).

3.2. EXPERIMENTAL SET-UP

The large difference between the resonant frequencies of electrons and nuclei, attributable to the relatively small mass of the former, for a given field, leads to the use of rather different electronic techniques in the two cases:

(a) Nuclear Magnetic Resonance (NMR)

One way of observing a nuclear resonance is to place the sample between the poles of a magnet and direct radiation at it at right-angles

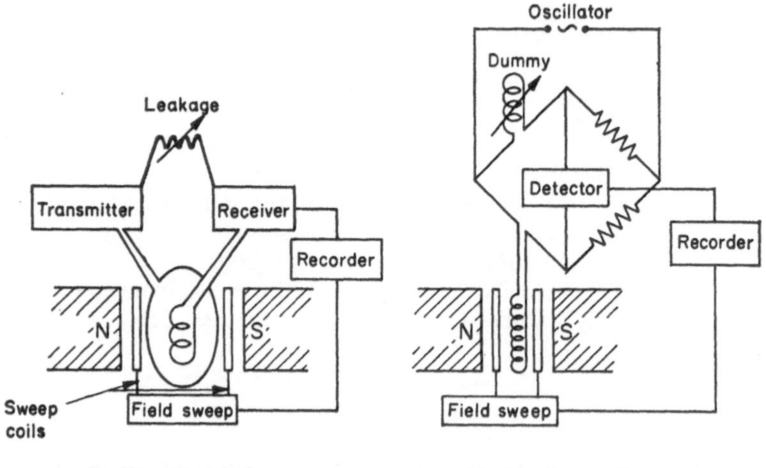

Figure 3-1. Schematic view of nuclear resonance experiment.

to the magnetic field from an appropriate radio transmitter. A radio-receiver is then placed at right-angles to the incident radiation (to avoid having to deal with the full power transmitted) in order to detect the dispersed signal (see Figure 3-1). At resonance the detected signal will be at maximum amplitude.

Another way of observing the resonance is by simply incorporating the sample in a tuned circuit so that it controls the inductance of an appropriate coil. Then under resonance conditions the a.c. power absorbed by the coil will be at a maximum so that we measure a decrease in the amplitude of the oscillations in that part of the circuit (i.e. compared with that in the "dummy" load).

Various modifications of these methods are used, including some of great ingenuity, such as the pulse techniques of Hahn in which a single coil is used both as the transmitter of a short intense pulse, and then as a receiver to detect "echoes" (see Pople, Bernstein and Schneider).

The resonance frequency is usually that of a stabilized crystal oscillator, which can be known to a high degree of accuracy. Resonance is reached by altering the field, by means of two subsidiary "sweep" coils. The main part of the field is commonly provided by either a permanent or an electromagnet.

It is difficult to get a field of sufficient homogeneity over the whole sample so liquid samples are often spun at a large enough rate to give effectively the same average applied field for all the nuclei; this sharpens the lines. Many other modifications are available for improving resolution and sensitivity (again see Pople *et al.*).

(b) Electron Spin Resonance (ESR)

Much larger variations in the field strength are required in an ESR spectrometer than in NMR and so, generally, permanent magnets are not suitable for producing the field.

The electronic arrangement is obviously very different from that in an NMR spectrometer mainly because the wavelength of the radiation is comparable with the physical dimensions of the various parts of the apparatus, e.g. X-band is 3 cm microwaves., This makes it relatively easy to direct the radiation from one place to another, by means of "waveguides" of appropriate dimensions, but the power sources (usually klystrons) are rather difficult to control accurately. However, the spectrometers used in electron resonance do not have to be constructed to the same degree of precision as those in NMR because the line-widths are 10^3–10^6 times greater.

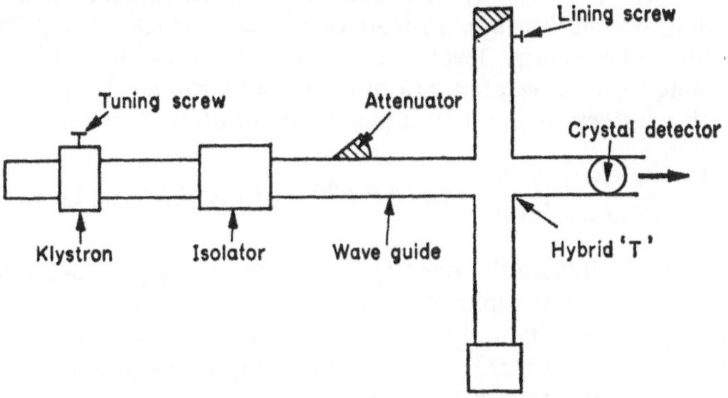

Figure 3-2. A bridge system used in an ESR spectrometer.

In Figure 3-2 we show the microwave equivalent to a bridge circuit. When the klystron is oscillating, standing electromagnetic waves are set up throughout the enclosed waveguide system. The two arms of the Hyrid T ("magic T"), one containing a dummy load, the other a cavity into which protrudes the sample, are of such dimensions that at "perfect tuning" virtually no microwave power is transmitted to the crystal detector. When resonance absorption is taking place the cavity is no longer matched to the dummy load, so power leaks to the crystal detector to be amplified and eventually displayed, either on a cathode ray tube or on a recorder synchronized with the field (see Ingrams).

3.3. PRELIMINARY LOOK AT THE INTENSITY OF ABSORPTION

The situation in an experiment is not as simple as it may seem from what we have discussed so far, because in a liquid or solid sample, each of a given resonating species will be in its own local field at any instant, consisting of the applied field and that due to its environment, i.e. other magnetic particles and charges in motion. If this local field were not fluctuating violently, due to rapid oscillations and Brownian motion, resonance would be observed over a wide range of applied fields and there would not be enough of the particles in any particular local field for us to observe any resonance effects at all. Luckily, however, the changes due to the random thermal motion in liquids, at ordinary temperatures, effectively average out the local fields to something not very different from that applied, providing the material is not ferromagnetic.

Another important consequence of the rapid field fluctuations at any point in a liquid, is that they can induce transitions between the magnetic energy levels of the various constituent particles. This means that, at equilibrium, instead of all the particles being in their lowest magnetic energy level, some will be in excited states. We can estimate the relative populations of two adjacent levels to a sufficient degree of accuracy, using a Boltzmann distribution function, i.e.

$$\frac{\text{number in lower level}}{\text{number in upper level}} = \exp\,(\Delta E/kT) = \exp\,(\gamma\hbar B/kT) \approx 1 + \gamma\hbar B/kT$$

where k is Boltzmann's constant and at ordinary temperatures, $kT \gg \gamma\hbar B$. The various magnetic energy levels are therefore very nearly equally populated with only a few more in the lower levels, e.g. for protons in fields of 10,000 gauss at 300 K, the above fraction is about 1.000007, i.e. $7/10^6$ more in lower level.

Since we can take the transition probability for absorption to be the same as that for emission for particles in the lower, upper states respectively, there will be a net absorption of the applied radiation, which will be proportional to the population difference. If we wished to increase the sensitivity of a magnetic resonance spectrometer we could do so by increasing the frequency, and hence the field for resonance, and/or by decreasing the temperature.

The total intensity of absorption and hence the sensitivity, depends on the inner properties of a sample as well as the number of the magnetic species present, for at the beginning of the irradiation of the sample the populations of the two levels will start evening-up, because more transitions occur away from the more densely populated states. What we would expect to observe is a transient absorption of power decreasing to zero as the populations of the various levels become equal, when the sample is said to be "saturated".

It is not difficult to examine this situation in simple mathematical terms:

Consider a particle, spin $\frac{1}{2}$, with two states labelled α and β. Let $W_{\alpha\beta}$, $W_{\beta\alpha}$ be the probabilities of transitions $\alpha\rightarrow\beta$ $\beta\rightarrow\alpha$ respectively in the absence of applied radiation, and n_α, n_β the numbers of particles with α, β spin. Then at equilibrium,

$$n_\alpha{}^0 W_{\alpha\beta} = n_\beta{}^0 W_{\beta\alpha} \tag{3.2}$$

If the system is disturbed from its equilibrium, the rate of return to steady conditions is given by:

$$d(n_\alpha - n_\beta)/dt = 2n_\beta W_{\beta\alpha} - 2n_\alpha W_{\alpha\beta} \tag{3.3}$$

Now if p is the probability of a transition induced by some applied radiation, Eq. (3.3) becomes:

$$d(n_\alpha - n_\beta)/dt = 2n_\beta(W_{\beta\alpha} + p) - 2n_\alpha(W_{\alpha\beta} + p)$$

i.e.

$$-d\Delta n/dt = (\Delta n - \Delta n^0)/\tau_1 + 2p\Delta n$$

where

$$\Delta n = n_\alpha - n_\beta \quad \text{and} \quad \tau_1 = (W_{\alpha\beta} + W_{\beta\alpha})^{-1}$$

$$= \text{spin–lattice relaxation time.}$$

Under steady conditions,

$$d\Delta n/dt = 0$$

i.e.

$$\Delta n = \Delta n^0/(1 + 2p\tau_1) \tag{3.4}$$

Eq. (3.3) is of great practical importance, for it tells us how to arrange the conditions of our experiments. On applying the radiation the population difference falls from Δn^0 to Δn and it is the latter which determines the intensity of the signal observed in a "slow-passage" magnetic resonance experiment. Thus we can see that if the radiation intensity is too much, p will be large compared with τ_1 and Δn will be small, i.e. the resonance absorption will be difficult to observe and easily saturated. Similarly, if the probability of a transition being induced by the thermal motions within the sample is small, τ_1 will be large and any steady resonance absorption small.

3.4. LINE SHAPES: CLASSICAL INTRODUCTION

For a given frequency there will be a spread of magnetic fields (i.e. applied) at which the magnetic resonance of an assembly of particles will be observed. This is because the thermal motion does not average out the local fields exactly, even when the applied field is strictly homogeneous. Generally, the slower the random movement of the particles, the less effectively as the field fluctuations averages out and the broader the resonance line will be (see Figure 3-3).

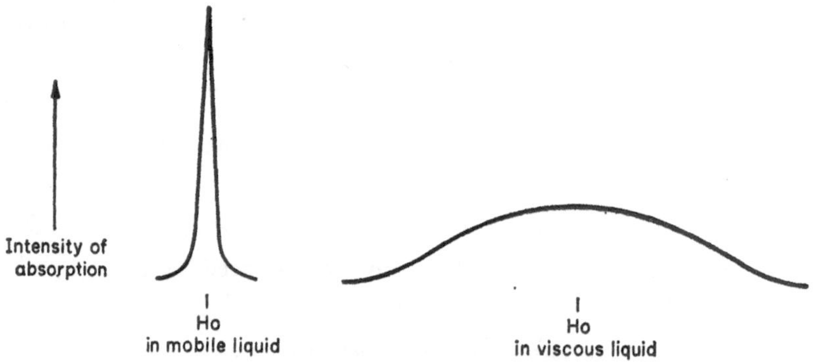

Figure 3-3. Line shapes in liquids.

We can examine the effects of these local field fluctuations in the following simplified classical way. When steady state conditions exist for a sample in a constant magnetic field B_z there will be an induced moment in the z direction due to the differential alignment of the spins as given by the Boltzmann distribution function. Let this induced moment be $M_z{}^0$. The components of this magnetization will be zero in the x and y directions because the individual spins will be precessing randomly, i.e. about their individual local fields.

If the system is now displaced from its equilibrium by some means or other, the rate of return of M_z to its steady value $M_z{}^0$, can be calculated from Eq. (3.3), i.e.

$$dM_z/dt = (M_z^0 - M_z)/\tau_1 \tag{a}$$

It is not unreasonable to assume that the other two components of the induced moment should also approach their steady values, i.e. zero, at a rate proportional to the displacement from equilibrium. Since these components have a symmetry, with respect to the applied field, which is different from that of M_z we assign a different proportionality constant τ_2, i.e.

$$dM_x/dt = -M_x/\tau_2 \tag{b}$$

$$dM_y/dt = -M_y/\tau_2 \tag{c}$$

The picture is however, still incomplete, because as soon as a magnetization appears in the xy plane, it will tend to precess about the z-axis, according to Eq. (1.6), so the equations for M_x and M_y will describe "damped" oscillations, i.e.

$$dM_x/dt = M_y B_z - M_x/\tau_2 \tag{d}$$

$$dM_y/dt = -M_x B_z - M_y/\tau_2 \tag{e}$$

i.e.

$$d(M_x^2 + M_y^2)/dt = -(M_x^2 + M_y^2)/\tau_2 \tag{f}$$

Hence from (a) and (f) we can verify that under steady conditions, i.e. when

$$dM_z/dt = d(M_x^2 + M_y^2)/dt = 0;$$

$$M_z = M_z^0, \qquad M_x^2 + M_y^2 = 0 = M_x = M_y$$

This process of returning to steady conditions is called "relaxation" and from (a) and (f) we see that τ_1, τ_2 govern the rates of re-establishing equilibrium values of the magnetization parallel and perpendicular to the applied field B_z, respectively.

If now a second magnetic field is applied, B' say, which can induce transitions between the levels of the individual particles we obtain the "Bloch" equations for the system, i.e.

$$\begin{aligned} dM_z/dt &= M_x B_y' - M_y B_x' - (M_z - M_z^0)/\tau_2 \\ dM_x/dt &= M_y B_z - M_z B_y' - M_x/\tau_2 \\ dM_y/dt &= M_z B_x' - M_x B_z - M_y/\tau_2 \end{aligned} \tag{3.5}$$

where we take $B_z' = 0$, since fields in the z-direction will not be effective in inducing transitions between levels with different z components of angular momentum (i.e. of the individual particles).

We could derive Eq. (3.5) in a rather more direct way, by first writing down the equation of motion for each of the individual spins, and then summing to obtain the resultant effect, i.e.

$$\Sigma d\bar{\mu}_i/dt = \Sigma\gamma\bar{\mu}_i \wedge \bar{B}_i$$

Now if the field at one of the particles is a constant, average, field $\bar{B}_0$ together with a fluctuating part $\Delta\bar{B}_i$, we can write the resultant equation of motion as:

$$d\bar{M}/dt = \gamma\bar{M} \wedge \bar{B}_0 + \Sigma\gamma\bar{\mu}_i \wedge \Delta\bar{B}_i \tag{3.6}$$

The Bloch equations (see Eq. (3.5)), then result from Eq. (3.6) when we make some assumptions about the last term.

Comparing the Bloch Eqs. (3.5) with (3.6), we can see that τ_1 is going to depend on the x and y components of the fluctuating field, and it is these which induce transitions between spin levels having different S_z values. τ_1, then, does not depend on the relative phases of the different spins. The situation is generally different to this for τ_2, because it depends on H_z (local), whose effect is not to induce transitions but to make the spins precess at their own individual rates, i.e. tending to randomize their phases. This is why τ_2 is called the "transverse" relaxation time. However, since it depends also on components of the field perpendicular to the z-axis, τ_2 will also depend on the rate at which transitions are induced; in fact, when the thermal motions of the particles in a liquid are fast enough, we can expect τ_1 to be the same as τ_2. In this situation, the different rates of precession due to different local fields no longer determine the rate at which the phases of the spins are randomized. (This is because the average local field becomes the same for all particles.)

3.5. SOLUTION OF THE BLOCH EQUATIONS FOR AN IRRADIATED SAMPLE

Suppose we apply linearly polarized electromagnetic radiation to our sample. Let the direction of the magnetic vector be that of the x-axis, so that its magnetic field is represented by $B_x = 2B' \cos \omega t$; $B_y = 0$, $B_z = 0$. We can resolve this field into two components, circularly polarized about the z-axis, as we do in the theory of optical rotation, i.e.

clockwise: $B_x = B' \cos \omega t;\quad B_y = -B' \sin \omega t$

(looking from $+z$ to $-z$ down on to the xy plane)

anti-clockwise: $B_x = B' \cos \omega t;\quad B_y = B' \sin \omega t$

Only one of these will be effective in causing a given spin to "flip" over, so we shall consider the clockwise component; then replacing γB_z by ω_0 we can write Eq. (3.5) as:

$$\begin{aligned} dM_z/dt &= -M_x\omega' \sin \omega t - M_y\omega' \cos \omega t - (M_z - M_z^0)/\tau_1 \\ dM_x/dt &= M_y\omega_0 + M_z\omega' \sin \omega t - M_x/\tau_2 \\ dM_y/dt &= -M_x\omega_0 + M_2\omega' \cos \omega t - M_y/\tau_2 \end{aligned} \quad (3.7)$$

where

$$\omega' = \gamma B'$$

The easiest way of solving these equations for the case when the resonance is obtained under steady conditions, is to change to a rotating coordinate system; i.e. put,

$$u = M_x \cos \omega t - M_y \sin \omega t$$

$$v = -M_x \sin \omega t - M_y \cos \omega t$$

It is easy to show that u is the component of the magnetization which is in-phase with the circularly-polarized radiation, and that v is $\frac{1}{2}\pi$ out-of-phase (see Figure 3-4).

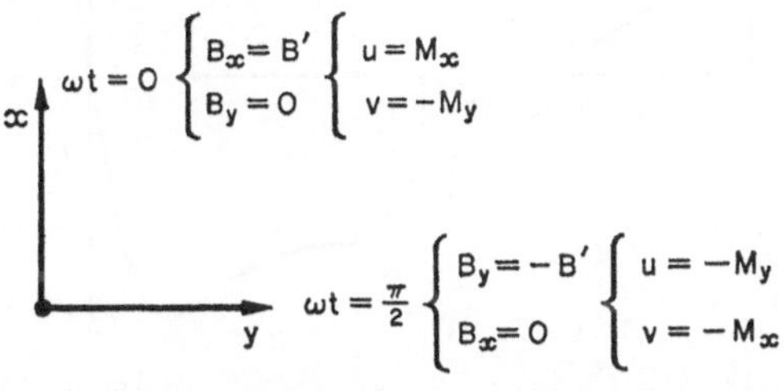

Figure 3-4. Relationship between radiation and "*u*" and "*v*".

In terms of these new coordinates Eq. (3.7) becomes:

$$\begin{aligned} dM_z/dt - \omega' v + (M_z - M_z^0)/\tau_1 &= 0 \\ du/dt + u/\tau_2 + (\omega_0 - \omega)\, v &= 0 \\ dv/dt + v/\tau_2 - (\omega_0 - \omega)\, u + \omega' M_z &= 0 \end{aligned} \quad (3.8)$$

While steady conditions are maintained (slow passage through signal) the induced moment along the line of the fixed field will be constant, i.e. $dM_z/dt = 0$, and the induced moment perpendicular to this will follow the rotating field, i.e. $du/dt = dv/dt = 0$. It is now relatively easy to solve for M_z, u and v, and we find, for example,

$$u = -\tau_2(\omega_0 - \omega)\, v = M_z^0 \frac{\omega' \tau_2^2(\omega_0 - \omega)}{1 + \tau_2^2(\omega_0 - \omega)^2 + \omega'^2 \tau_1 \tau_2} \tag{3.9}$$

Now the torque exerted on the induced moment by the radiation is $(M_x B_y - M_y B_x)$, so the rate at which the radiation does work, i.e. loses energy, is angular velocity × torque, i.e.

$$\begin{aligned} \text{rate of energy absorption} &= \omega M_x(-B' \sin \omega t) - \omega M_y(B' \cos \omega t) \\ &= \omega B' v \\ &= \frac{\omega B' M_z^0 \omega' \tau_2}{1 + \tau_2^2(\omega_0 - \omega)^2 + \omega'^2 \tau_1 \tau_2} \end{aligned}$$

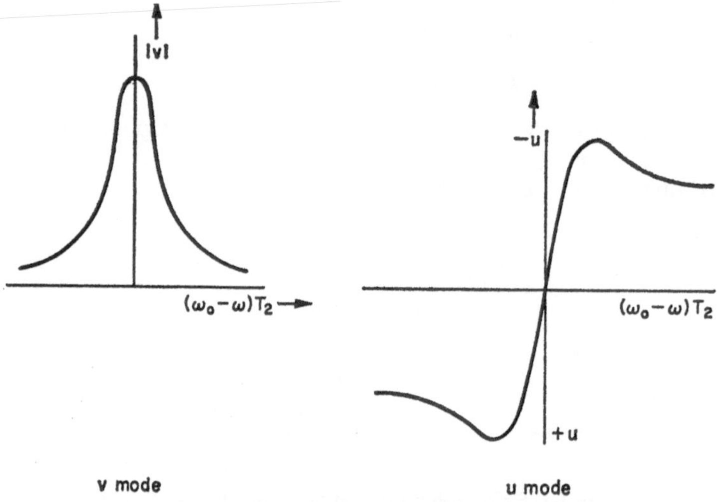

Figure 3-5. Variation of u and v with irradiation frequency.

So at low power the line shape is Lorentzian, given by the function

$$f(\omega - \omega_0) = \frac{k}{1 + \tau_2^2(\omega - \omega_0)^2} \tag{3.10}$$

k = constant for the given conditions.

This is why v is called the "absorption" mode. u is the "dispersion" mode and the variation of the two modes with the frequency, $\omega/2\pi$, of the applied radiation is shown graphically in Figure 3-5.

In an actual experiment we are detecting some variation in the magnetization in the x- or y-directions, and are also generally using linearly polarized radiation. What we measure is therefore going to be some mixture of the absorption and dispersion modes, i.e. v and u. In order to measure only one of these it is necessary to cancel out the effects of the other in the detected signal. In the single-coil method this is done by varying the impedance of the dummy load and in the double-coil method by mixing a small fraction of the input signal with that received, by some means or another, (see Pople, Bernstein and Schneider). These are shown schematically in Figure 2-1.

This completes our treatment of the general, "background" theory on which the phenomenon of magnetic resonance is based. In the following chapters we shall discuss how the effects observed in specific types of experiment, on specific systems, can to a certain extent be rationalized in terms of these general principles.

Chapter 4

Parameters obtained from Magnetic Resonance Spectra

4.1. INTRODUCTION

Magnetic resonance spectra are generally recorded on a paper chart, changing conditions in the microwave circuits being converted to deflections from the base line in such a way that the extra power absorbed due to resonance is measured by the amplitude of the deflection. Ideally the chart moves past the pen at a rate linearly geared to the field, so that the intensity of the absorption is proportional to the area between the extrapolated base line and the curve, which has a Lorentzian or Gaussian shape, depending on the relaxation conditions in the sample (see Figure 4-1).

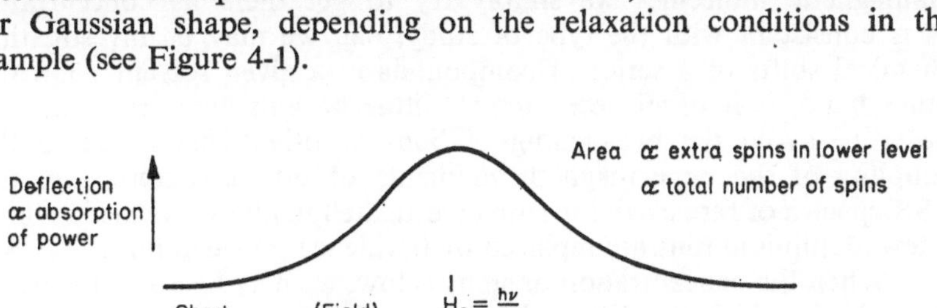

Figure 4-1. Recording of a resonance.

A magnetic resonance experiment can be classified into two parts: first, we have to arrange conditions so that we can observe a spectrum of appropriate quality, and secondly, we have to measure the positions and shapes of the lines. From the line-shapes we can deduce relaxation times, lifetimes of states or molecules and so on; we shall discuss such time-dependent phenomena in a later chapter. From the positions of the lines we can evaluate the magnetic resonance parameters of the sample; e.g. chemical or *g*-shifts, coupling constants, etc.

Experimentally the chemical shift of a nucleus, and the *g*-value of an odd electron system, define the positions of the "centres" of corresponding resonances; i.e. at what fields and frequencies the resonances would occur if there were no other magnetic particles in the system.

In NMR these positions are related to a standard resonance and the chemical shift is the ratio of the difference in resonant fields (at constant frequency) between the sample and the standard, divided by the resonant field of the standard, multiplied by 10^6 (hence it is given in p.p.m. = parts per million). With this definition up-field shifts (diamagnetic) are positive and this is what we shall adopt, although some people prefer to define things in terms of constant fields and call "up-frequency" shifts positive instead.

Coupling constants define the separations of the lines in the resonance spectrum of a particular magnetic species, at least to the first order of approximation, and do not depend on the applied field. They arise from the interactions of the magnetic particles in the sample with each other and are given in frequency or in field units.

Arranging Experimental Conditions

Generally speaking we wish to obtain a spectrum which is as intense as possible, without sacrificing resolution, and the first variable we can try to optimize is the concentration of the spins. For nuclei in diamagnetic molecules we simply try to get them as concentrated as is consistent with the type of study, e.g. we may be investigating chemical shifts of a series of compounds in a given solvent. On the other hand, it is often necessary to dilute paramagnetic molecules in order to avoid the broadening of lines or other effects due to the coupling of the large magnetic moments of odd electrons; e.g. the ESR spectra of rare earth ions are investigated in alum crystals in which a few aluminium ions are replaced by trivalent rare earth ions.

When the concentration of spins is low, we may be able to concentrate the material, e.g. C^{13} enrichment, or we may employ a computer (CAT) to record and add together many spectra over the same range. Since the noise is random, these additions will tend to even it out. On the other hand, the resonance lines always occur in the same place so they always add to the corresponding lines in the previous "run", i.e. the amplitude of the lines increases relative to the noise after each addition. Unfortunately a CAT is expensive, and its use implies a very good field stabilization and the availability of a spectrometer for a number of hours, since the spectrum has to be scanned perhaps a large number of times.

4.2. PARAMETERS FROM LIQUID SAMPLES

Magnetic resonance spectra obtained from liquid samples are usually relatively simple, because anisotropic magnetic interactions are effectively averaged out by the rapid molecular tumbling. This means that we get, therefore, only one *g*-value for a radical in its ESR spectrum, one chemical shift for each type of nucleus in NMR spectra, and one coupling constant for each pair of interacting magnetic dipoles.

The *g*-value of a free radical or transition metal ion in solution, can be measured either directly, if we know the resonant frequency and can measure the field with sufficient accuracy, say by means of a proton resonance probe, or indirectly, i.e. by measuring the difference in field between the centre of resonance and that of some standard radical, e.g. potassium nitroxylate.

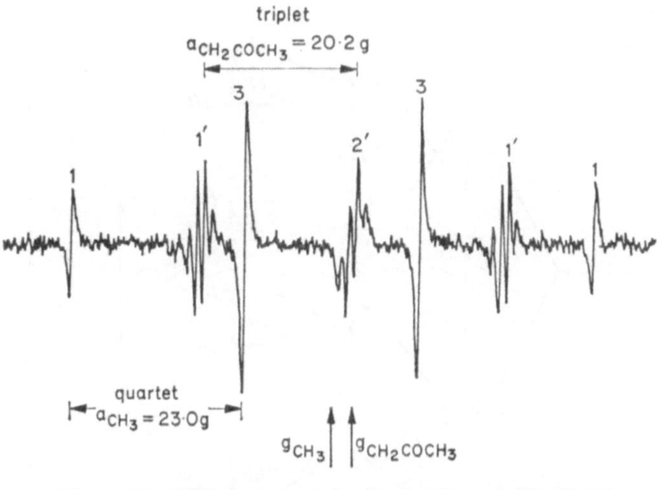

Figure 4-2. ESR of mixture of radicals CH_3 and CH_2COCH_3.

Whereas the *g*-value is appropriate for defining the centre of an ESR spectrum, because it determines the magnetogyric ratio of an odd electron, in NMR the so-called chemical shift, σ, is employed, and is defined by the equation:

$$h\nu = \gamma_N \hbar(1-\sigma)\, B = g_N \beta_N (1-\sigma)\, B \tag{4.1}$$

γ_N the nuclear magnetogyric ratio, and g_N are constant for a given nucleus, but its resonance occurs at different applied fields, *B*, in different chemical environments. σ is usually quoted in parts per million and is referred to the nucleus in some chosen compound rather than to the bare nucleus. The chemical shift is the fractional change in field going from the standard to the centre of the spectrum being examined, and is positive when the change in applied field is positive; i.e. when the nucleus is more heavily "shielded".

The most commonly used scale for protons is the τ (tau) scale. To find the tau-value you simply find the chemical shift relative to that in tetramethylsilane, $Si(CH_3)_4$, and add 10.

When there is more than one nucleus of a given isotope in a molecule, it is often difficult to determine their relative chemical shifts. The complications which arise are only significant when the differences in chemical shifts are comparable with the indirect coupling of the nuclear spins. We shall discuss them after the next section on coupling constants.

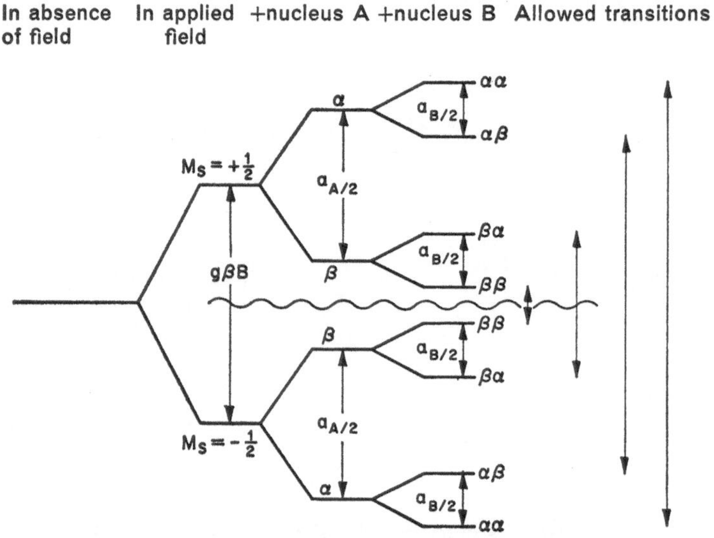

Figure 4-3. Electron spin energy levels with two nearby protons.

Coupling Constants

The field experienced by a resonating spin consists of the applied field plus the modifying effect of the Larmor precession of all the electrons and the small fields due to other magnetic particles. The interactions with these small fields may be direct, i.e. dipole–dipole, or indirect, arising from the spin polarization of the paired electrons. The effective field at the resonating spin depends on the orientations of these other spins, so that the single absorption which would be observed in the absence of other nuclear magnets becomes split into $(2I+1)$

lines for each nuclear spin *I*. This is best seen diagrammatically and we take as an example the case of an odd electron interacting with two nuclei spin $\frac{1}{2}$ (see figure 4-3).

The ESR spectrum observed for this system would be four lines of equal amplitudes separated by the coupling constants (in field units, gauss), for example see Figure 4-4.

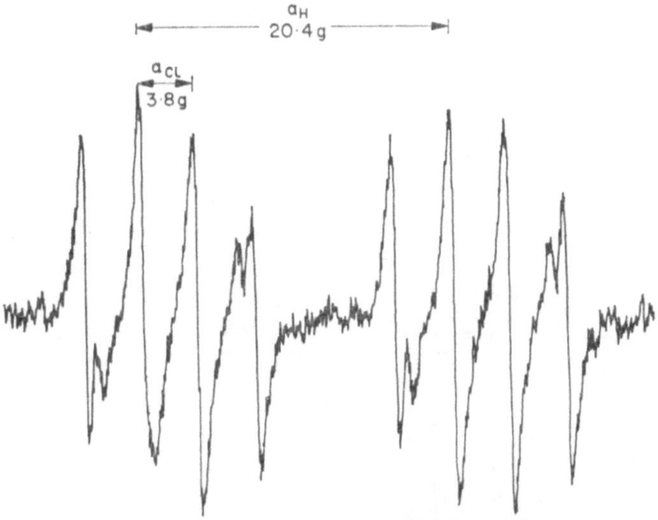

Figure 4-6. ESR spectrum of CHCl COOH

It is interesting to note at this point that ESR spectra are usually presented in the derivative form, since this makes it easier to identify the lines and measure their relative positions. Frequently the absorption band is even differentiated twice before presentation on the chart recorder.

When two or more of the nuclei are equivalent, some lines will overlap and the relative intensities change accordingly, e.g. two equivalent protons give a 1 : 2 : 1 three-line ESR spectrum, three equivalent protons lead to a 1 : 3 : 3 : 1 splitting, four, a 1 : 4 : 6 : 4 : 1 splitting, and so on, i.e. the intensities are in the ratios of binomial coefficients. Figure 4-5 illustrates how spectra are split up by sets of equivalent protons.

Similar rules apply for the effects of nuclei which have spins greater than $\frac{1}{2}$, for example, N^{14} has a spin of 1 and therefore splits a resonance into 3 lines, Cl^{35} and Cl^{37} both have a spin of 3/2 and therefore multiply the number of lines by 4 (see Figure 4-6).

Most magnetic resonance spectra are much more complex than those illustrated above due to the relatively large number of interacting

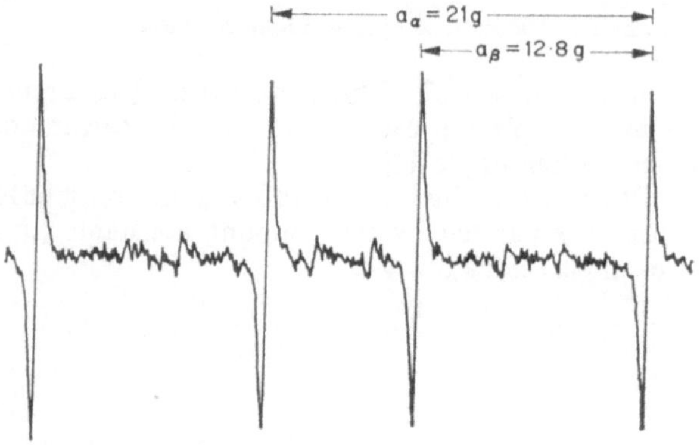

Figure 4-4. ESR spectrum of $\dot{C}_{\alpha}HCOOH\ C_{\beta}H(OH)COOH$.

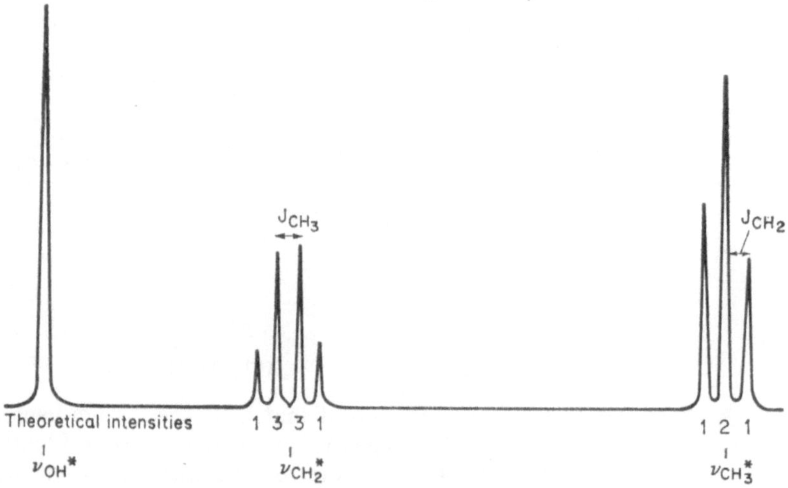

Figure 4-5(a). NMR spectrum of ethanol (with 2% dil HCl).

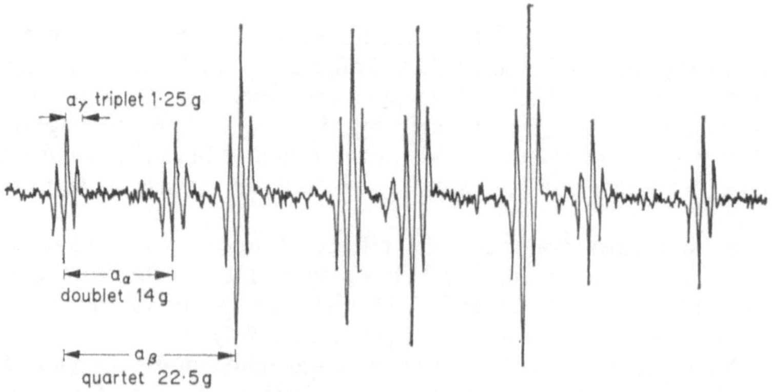

Figure 4-5(b). ESR spectrum of $\underset{\beta}{CH_3}\,\underset{\alpha}{\dot{C}HO}\underset{\gamma}{CH_2}\,CH_3$.

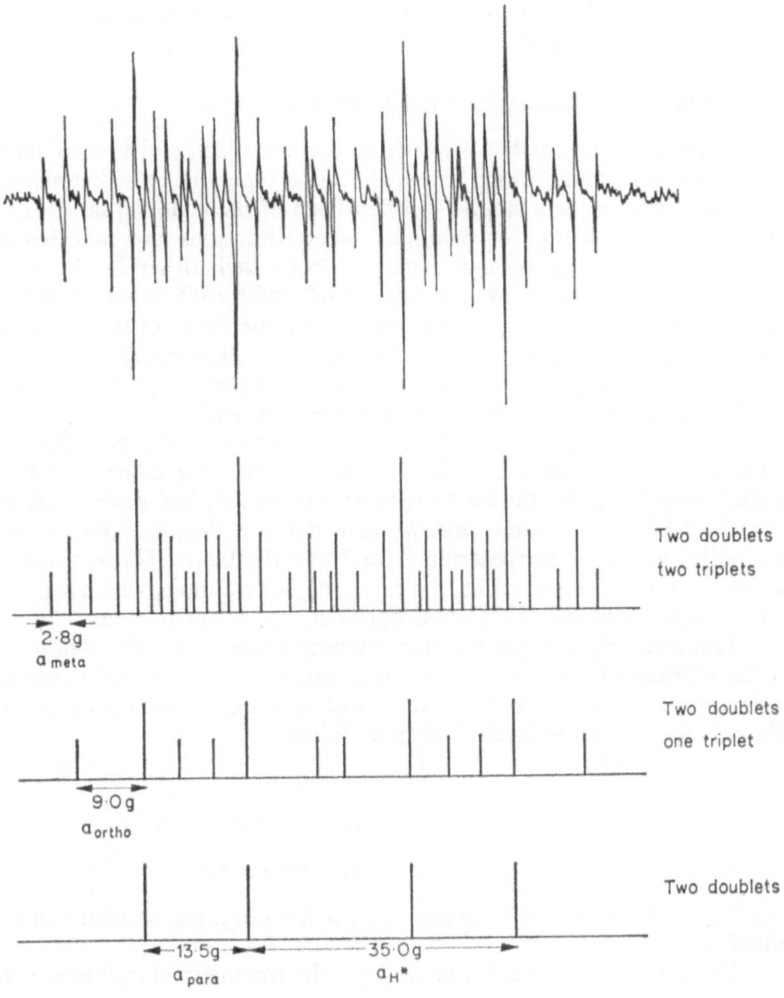

Figure 4-7. ESR spectrum of 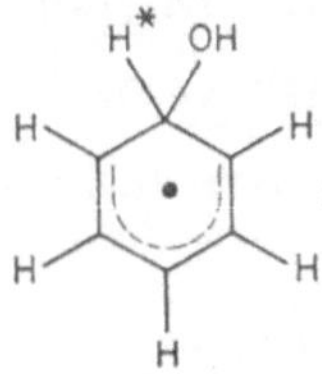

radical and its breakdown.

spins, but the spectra can be thought of in the same way, i.e. as being built from successive multiplication of lines, taking account of overlapping where it is appropriate; e.g. see Figure 4-7.

4.3. SOME SECOND-ORDER EFFECTS

When we look at the magnetic resonance spectra of two or more non-interacting spins, which are of the same type and in rather similar environments, we see a simple superposition of their individual spectra. This situation occurs, for example, when the spins are in different molecules or are far enough removed from each other in the same molecule. If the spins interact appreciably with each other, however, the situation becomes more complicated, and the shape of the spectrum depends on the relative values of their interaction energy (coupling constant) and the difference between their resonant frequencies, which are determined by their chemical shifts and the field.

We can see how this complication arises if we consider two identical protons (in this context this means that they are magnetically equivalent). According to the first-order theory which we have used till now, in which we assume that we can assign a definite spin to each magnetic particle, their spectrum should be a doublet. This is contrary to what we have observed in Figure 4-5(a), for example, where it is evident that no splitting is observed between equivalent protons.

This dilemma is resolved when we admit that under the conditions of the experiment, the protons are indistinguishable and that therefore their combined spin wave-functions must take this factor into account. The scheme of energy levels is then as follows:

$\alpha\alpha$	$h(\nu+J/4)$
$\beta\beta$	$h(-\nu+J/4)$
$(\alpha\beta\pm\beta\alpha)/\sqrt{2}$	$h(-J/4\pm J/2)$

where ν is the resonant frequency and J the coupling constant of the nuclei.

These energies are calculated using the magnetic Hamiltonian for the system, i.e.

$$\mathcal{H}=\tfrac{1}{2}h\nu(I_A+I_B)+hJI_A.I_B$$

and the energy is simply $\langle\mathcal{H}\rangle$.

The energy of the state $(\alpha\beta+\beta\alpha)/\sqrt{2}$ is calculated, for example, in the following way:

$$\tfrac{1}{2}\langle\alpha\beta+\beta\alpha\mid\mathcal{H}\mid\alpha\beta+\beta\alpha\rangle=$$
$$\tfrac{1}{2}hJ\langle\alpha\beta+\beta\alpha\mid I_{Az}I_{Bz}+I_{Ax}I_{Bx}+I_{Ay}I_{By}\mid\alpha\beta+\beta\alpha\rangle$$

i.e.

$$E_{(\alpha\beta+\beta\alpha)/\sqrt{2}} = -\tfrac{1}{4}hJ + \tfrac{1}{4}hJ\langle\alpha\beta+\beta\alpha \mid I_{A+}I_{B-}+I_{A-}I_{B+} \mid \alpha\beta+\beta\alpha\rangle \quad (4.2)$$

where I_{A+}, I_{A-}, etc. are the "raising" and "lowering" operators and are defined as follows: for a spin "I" (also see Appendix):

$$I_+ = I_x + iI_y, \qquad I_- = I_x + iI_y$$

The properties of these operators are found from the commutation relations between the components of I, i.e.

$$iI_z = I_xI_y - I_yI_x \qquad \text{etc.}$$

For the raising operator I_+:

$$I_zI_+ = I_+I_z + I_+$$

If m_z is the eigenvalue of I_z associated with the eigenfunction $\mid m_z\rangle$ we find

$$I_z(I_+ \mid m_z\rangle) = (m_z+1)\,(I_+ \mid m_z\rangle)$$

Hence $I_+ \mid m_z\rangle$ is proportional to the function $\mid m_z+1\rangle$ provided that m_z is not the highest eigenvalue of I_z, i.e. application of I_+ to an eigenfunction of I_z changes it effectively into the eigenfunction with the next highest eigenvalue similarly $I_- \mid m_z\rangle \rightarrow \mid m_z-1\rangle$.

For particles with spin $\tfrac{1}{2}$ the relevant relationships are:

$$\begin{cases} I_+ \mid \alpha\rangle = I_- \mid \beta\rangle = 0 \\ I_+ \mid \beta\rangle = \mid \alpha\rangle \\ I_- \mid \alpha\rangle = \mid \beta\rangle \end{cases} \quad (4.3)$$

Using these relationships, Eq. (4.2), for the energy becomes:

$$E_{(\alpha\beta+\beta\alpha)/\sqrt{2}} = -\tfrac{1}{4}hJ + \tfrac{1}{4}hJ\langle\alpha\beta+\beta\alpha \mid \beta\alpha+\alpha\beta\rangle$$

$$= +\tfrac{1}{4}hJ$$

For transitions to occur between two states they have to have the same symmetry and their spins must differ by one unit of spin—hence only one energy change is possible, i.e. $h\nu$.

In practice the situation is often somewhere between the extreme cases which are labelled the A_2 or AA case, when the two nuclei are equivalent, and the AX case, when the difference between their chemical shifts is large enough to lead to a spectrum of four lines of equal amplitudes, resolution permitting. Intermediate cases are labelled AB.

The AB Type of Spectrum

As we have just implied, when there is appreciable interaction between two nuclei, simple products, such as $\alpha\beta$ or $\beta\alpha$, are no longer adequate approximations to the true wavefunctions. In order to obtain better approximations we take linear combinations of these products. Derived from the four products $\alpha\alpha$, $\beta\beta$, $\alpha\beta$ and $\beta\alpha$ we obtain four first-order wavefunctions, i.e.

$$\begin{array}{l} \alpha\alpha \\ \beta\beta \\ c_1\alpha\beta + c_2\beta\alpha \\ c_2\alpha\beta - c_1\beta\alpha \end{array}$$

The coefficients are determined using the variational principle in conjunction with the magnetic part of the Hamiltonian. The secular determinant which has to be solved is:

$$\begin{vmatrix} \langle\alpha\beta \mid \mathscr{H} \mid \beta\alpha\rangle - E & \langle\alpha\beta \mid \mathscr{H} \mid \beta\alpha\rangle \\ \langle\beta\alpha \mid \mathscr{H} \mid \beta\alpha\rangle & \langle\beta\alpha \mid \mathscr{H} \mid \alpha\beta\rangle - E \end{vmatrix} = 0 \tag{4.4}$$

In frequency units the magnetic Hamiltonian is

$$\mathscr{H} = \nu_A I_{Az} + \nu_B I_{Bz} + J(I_{Az}I_{Bz} + I_{Ax}I_{Bx} + I_{Ay}I_{By}) \tag{4.5}$$

To work out the matrix elements we need the following:

$$\begin{aligned} \mathscr{H} \mid \alpha\alpha\rangle &= (+\tfrac{1}{2}\nu_A + \tfrac{1}{2}\nu_B + \tfrac{1}{4}J) \mid \alpha\alpha\rangle \\ \mathscr{H} \mid \beta\beta\rangle &= (-\tfrac{1}{2}\nu_A - \tfrac{1}{2}\nu_B + \tfrac{1}{4}J) \mid \beta\beta\rangle \\ \mathscr{H} \mid \alpha\beta\rangle &= (+\tfrac{1}{2}\nu_A - \tfrac{1}{2}\nu_B - \tfrac{1}{4}J) \mid \alpha\beta\rangle + J/2 \mid \beta\alpha\rangle \\ \mathscr{H} \mid \beta\alpha\rangle &= (-\tfrac{1}{2}\nu_A + \tfrac{1}{2}\nu_B - \tfrac{1}{4}J) \mid \beta\alpha\rangle + J/2 \mid \alpha\beta\rangle \end{aligned}$$

To obtain the last two equations we have used the identity

$$I_{Ax}I_{Bx} + I_{Ay}I_{By} \equiv \tfrac{1}{2}(I_{A+}I_{B-} + I_{A-}I_{B+})$$

The equation for the two energy levels for which zeroth-order wavefunctions are not good enough is Eq. (4.4) which becomes:

$$\begin{vmatrix} (\tfrac{1}{2}\nu_A - \tfrac{1}{2}\nu_B - \tfrac{1}{4}J - E) & J/2 \\ J/2 & (-\tfrac{1}{2}\nu_A + \tfrac{1}{2}\nu_B - \tfrac{1}{4}J - E) \end{vmatrix} = 0$$

i.e.

$$E = -\tfrac{1}{4}J \pm \tfrac{1}{2}\sqrt{(\nu_A - \nu_B)^2 + J^2}$$

The other two levels have energies

$$\pm\tfrac{1}{2}(\nu_A+\nu_B)+\tfrac{1}{4}J$$

Transition Probabilities

In a magnetic resonance experiment we subject the sample to a fluctuating magnetic field polarized in a direction perpendicular to the steady applied field. If we let the direction of polarization be the x axis, then the probability per unit time of a transition between two states, 1 and 2 say, is from section 2.5, proportional to

$$|\langle 1 \mid \mathscr{H}'(t) \mid 2\rangle|^2 = |\langle 1 \mid B'_x(I_{Ax}+I_{Bx})\,\gamma_N \mid 2\rangle|^2$$

i.e. proportional to $|\langle 1 \mid (I_{A+}+I_{A-}+I_{B+}+I_{B-}) \mid 2\rangle|^2$.

The relative transition probabilities between the four states of the AB system are therefore:

Transition	Relative Intensity (α Transition Probability)	[ ]
$\alpha\alpha \longleftrightarrow \beta\beta$	0	$\nu_A+\nu_B$
$c_1\alpha\beta+c_2\beta\alpha \longleftrightarrow c_2\alpha\beta-c_1\beta\alpha$	0	$\sqrt{(\nu_A-\nu_B)^2+J^2}$
$\alpha\alpha \longleftrightarrow c_1\alpha\beta+c_2\beta\alpha$	$(c_1+c_2)^2$	$\frac{1}{2}(\nu_A+\nu_B)+\frac{1}{2}J-\frac{1}{2}\sqrt{(\nu_A-\nu_B)^2+J^2}$
$\alpha\alpha \longleftrightarrow c_2\alpha\beta-c_1\beta\alpha$	$(c_1-c_2)^2$	$\frac{1}{2}(\nu_A+\nu_B)+\frac{1}{2}J+\frac{1}{2}\sqrt{(\nu_A-\nu_B)^2+J^2}$
$\beta\beta \longleftrightarrow c_1\alpha\beta+c_2\beta\alpha$	$(c_1+c_2)^2$	$\frac{1}{2}(\nu_A+\nu_B)-\frac{1}{2}J+\frac{1}{2}\sqrt{(\nu_A-\nu_B)^2+J^2}$
$\beta\beta \longleftrightarrow c_2\alpha\beta-c_1\beta\alpha$	$(c_1-c_2)^2$	$\frac{1}{2}(\nu_A+\nu_B)-\frac{1}{2}J-\frac{1}{2}\sqrt{(\nu_A-\nu_B)^2+J^2}$

The observed spectrum is therefore of four lines symmetrically placed about the mean chemical shift (given by $\frac{1}{2}(\nu_A+\nu_B)$). Back substitution into the secular equations gives the coefficients c_1 and c_2 and after some algebra an interesting relation is found, i.e.

intensity of transition × distance from centre of the spectrum = constant

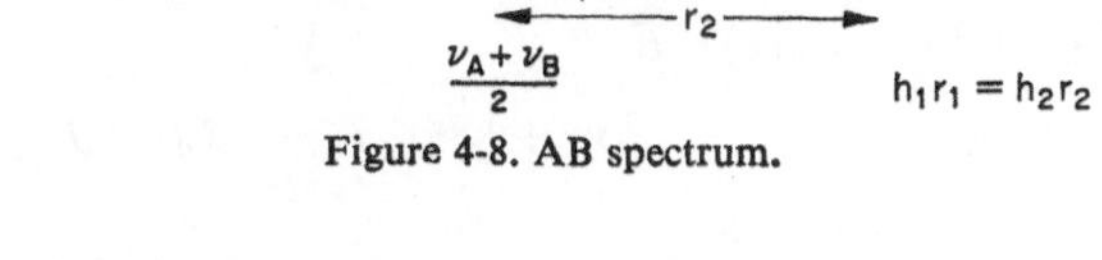

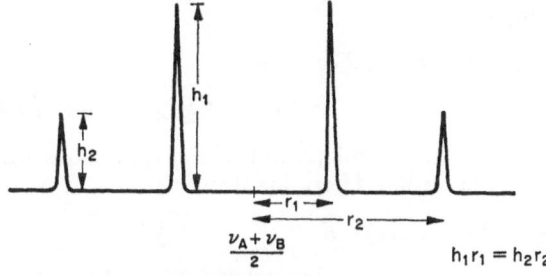

Figure 4-8. AB spectrum.

The two extreme cases are (1) when the chemical shift difference is large compared with the coupling between the two nuclei, i.e. $|\nu_A - \nu_B| \gg J$ when there are four lines of approximately equal intensity (difference in r_1, r_2 small compared with r_1). (2) when the coupling is large compared with the chemical shift, i.e. $|\nu_A - \nu_B| \ll J$ when $c_1 = c_2$ the outer lines have zero intensity and the inner lines are fused into one.

We shall now illustrate further how the shape of an NMR spectrum is influenced by second-order effects. We shall consider the case when two chemically equivalent nuclei A and A' couple unequally with a third nucleus X, which may be of the same isotope, as in the following ^{13}C enriched xylene

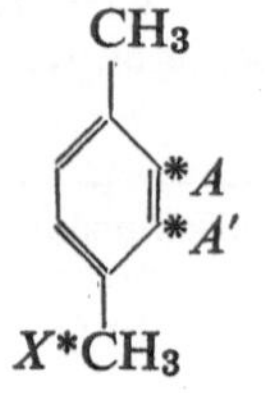

or a different nucleus altogether, e.g.

$$\underset{AX}{H^{13}C} \equiv \underset{A'}{CH}$$

We designate this the $AA'X$ case, and the magnetic Hamiltonian is:

$$\mathscr{H} = \nu_A(I_{Az} + I_{A'z}) + \nu_X I_{Xz} + J_{AA'} I_A . I_{A'} + J_{AX} I_A . I_X + J_{A'X} I_{A'} . I_X$$

ν_A, ν_X are the resonant frequencies of the nuclei at that field and the energy is measured in units of h.

The method is then, to take linear combinations of the zeroth-order spin functions, such as $\alpha\alpha\alpha$, $\alpha\beta\alpha$, $\beta\alpha\beta$, etc., and apply the variational principle to obtain a better description of the energy levels. The only complication comes from the states in which A and A' have opposite spins, for example:

$$\mathscr{H}\,|\,\alpha\beta\alpha\rangle = [\tfrac{1}{2}\nu_X + \tfrac{1}{4}(-J_{AA'} + J_{AX} - J_{A'X})] \times |\,\alpha\beta\alpha\rangle + \tfrac{1}{2}J_{AA'}\,|\,\beta\alpha\alpha\rangle + \tfrac{1}{2}J_{A'X}\,|\,\alpha\alpha\beta\rangle$$

In order to find the best combination of the spin states $\alpha\beta\alpha$ and $\beta\alpha\alpha$ we have to solve the secular determinant:

$$\begin{vmatrix} \tfrac{1}{2}\nu_X + \tfrac{1}{4}(-J_{AA'} + J_{AX} - J_{A'X}) - E & \tfrac{1}{2}J_{AA'} \\ \tfrac{1}{2}J_{AA'} & \tfrac{1}{2}\nu_X + \tfrac{1}{4}J(-J_{AA'} - J_{AX} + J_{A'X}) - E \end{vmatrix} = 0$$

The energy levels are (again in frequency units):

$$\left.\begin{matrix}\alpha\alpha\alpha \\ \beta\beta\beta\end{matrix}\right\} (\nu_A+\nu_X/2)+\tfrac{1}{4}(J_{AA'}+J_{AX}+J_{A'X})$$

$$\left.\begin{matrix}\alpha\alpha\beta \\ \beta\beta\alpha\end{matrix}\right\} \pm(\nu_A-\nu_X/2)+\tfrac{1}{4}(J_{AA'}-J_{AX}-J_{A'X})$$

$$\left.\begin{matrix}c_1\alpha\beta\alpha+c_2\beta\alpha\alpha \\ c_2\alpha\beta\alpha-c_1\beta\alpha\alpha\end{matrix}\right\} +\tfrac{1}{2}\nu_X-\tfrac{1}{4}(J_{AA'}\pm\sqrt{(J_{AX}-J_{A'X})^2+4J_{AA'}{}^2})$$

$$\left.\begin{matrix}c_1\beta\alpha\beta+c_2\alpha\beta\beta \\ c_2\beta\alpha\beta-c_1\alpha\beta\beta\end{matrix}\right\} -\tfrac{1}{2}\nu_X-\tfrac{1}{4}(J_{AA'}\pm\sqrt{(J_{AX}-J_{A'X})^2+4J_{AA'}{}^2})$$

where $\qquad c_2c_1=\sqrt{1+\lambda^2}+\lambda,\ \lambda=(J_{AX}-J_{A'X}/2J_{AA'}$

Let us look first at the X part of the spectrum, i.e. lines arising from changes in the state of X. The transition probabilities and therefore the intensities depend, as before, on the term

$$|\langle 1\mid I_{X+}+I_{X-}\mid 2\rangle|^2$$

We can now write down all of the transitions in the X part of the spectrum with the corresponding intensities:

$$\nu_X+\tfrac{1}{2}(J_{AX}+J_{A'X})\ldots 1$$
$$\nu_X-\tfrac{1}{2}(J_{AX}-J_{A'X})\ldots 1$$
$$\nu_X\ldots 4c_1{}^2(1-c_1{}^2)$$
$$\nu_X+\sqrt{1+\lambda^2})\ldots(1-2c_1{}^2)^2$$
$$\nu_X-\sqrt{1+\lambda^2})\ldots(1-2c_1{}^2)^2$$

The spectrum should, then, look like Figure 4-9.

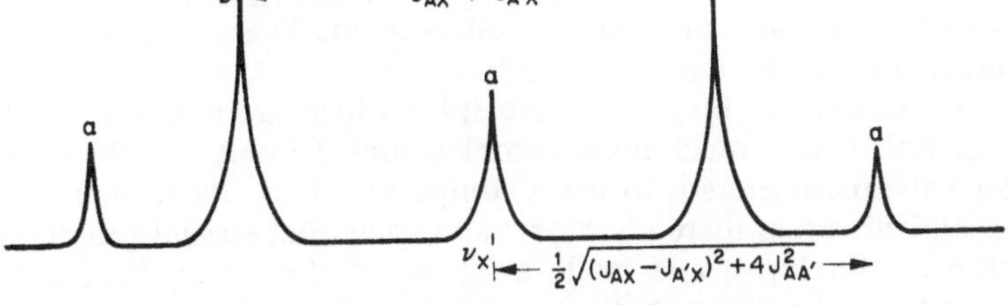

Figure 4-9. X part of $AA'X$ spectrum.

It is interesting to examine the limiting cases, first, when the A nuclei are magnetically equivalent, $J_{AX}=J_{A'X}$ and $\lambda=0$. The intensity of the outer "a" lines will be zero and that of the centre line will have a relative intensity of 2, i.e. the spectrum will be a 1 : 2 : 1 triplet, splitting J_{AX}.

Secondly, when A and A' are only weakly coupled together, $J_{AA'} \ll | J_{AX}-J_{A'X} |$, λ will be large and the central line will have approximately zero intensity. The spectrum will then be four lines of equal intensities. The splittings observed will then be J_{AX} and $J_{A'X}$.

The AA' part of the spectrum can be approached in a similar way and it consists of two quartets, starred and unstarred in Figure 4-10, separated by the average of J_{AX} and $J_{A'X}$. Four of the lines have intensities $1+1/\sqrt{1+\lambda^2}$, and the other four $1-1/\sqrt{1+\lambda^2}$.

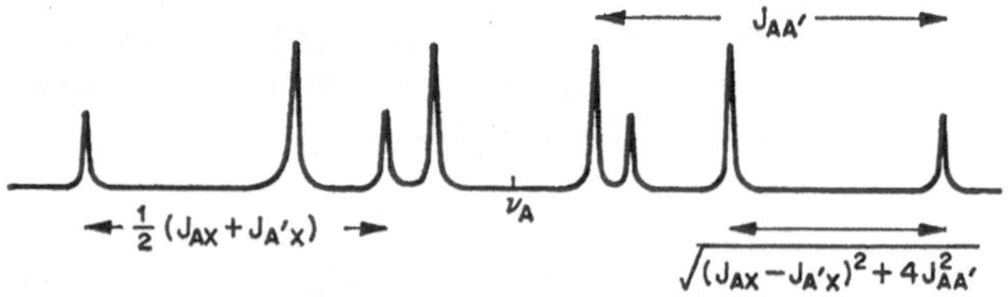

Figure 4-10. "A" part of $AA'X$ spectrum.

In the limit, when A and A' become exactly equivalent, the four lines of smaller intensity disappear, and the others coalesce so that there are left only two lines separated by J_{AX}. On the other hand, when $J_{AA'}$ is small compared with $J_{AX}-J_{A'X}$, all eight lines have equal intensities and the spectrum is simply the overlapping of the individual spectra of A and A', i.e. as if they were in different molecules. The splittings observed are then J_{AX} and $J_{AA'}$ in one quartet, and $J_{A'X}$ and $J_{AA'}$ in the other.

By comparing the X- and AA'-parts of the spectrum, it is sometimes possible to determine the relative signs of J_{AX} and $J_{A'X}$, i.e. by consideration of the splittings and the intensities of the lines, according to the formulae derived above.

In cases where there are more spins which are nearly equivalent, the analysis is that much more complex, and the only practical way of finding the parameters is to use a computer. Even then, the situation is so difficult when there are 5 or more spins that special programmes have been developed and made available to those who have not the time or know-how to write their own.

4.4. MAGNETIC RESONANCE IN SOLID SAMPLES

Leaving aside relaxation effects, the main difficulty when we are dealing with solid samples is that the g-values, chemical shifts and coupling constants are all dependent on orientation, and are not effectively averaged out by molecular motion. We shall consider the g-value of an odd-electron molecule, say a free radical or transition metal ion, in order to indicate how to set about obtaining the appropriate parameters from the magnetic resonance spectrum of a solid sample.

First, we should rewrite the equation for the magnetic energy in an applied field, i.e.

$$\mathscr{H} = \beta\bar{B}.\bar{\bar{g}}.\bar{S}$$
$$= \beta(B_z g_{zz} S_z + B_z g_{zx} S_x + B_z g_{zy} S_y \ldots)$$

$\bar{\bar{g}}$ here is no longer a simple number but a tensor of the second rank, having nine components, and it expresses the effects of spin–orbit coupling on the "true" spin. $\bar{S}$ is not the true spin but an "effective" or "fictitious" spin, containing a certain element of the orbital angular momentum (see the next chapter). Its components still have eigenvalues of $\pm\frac{1}{2}\hbar$, however.

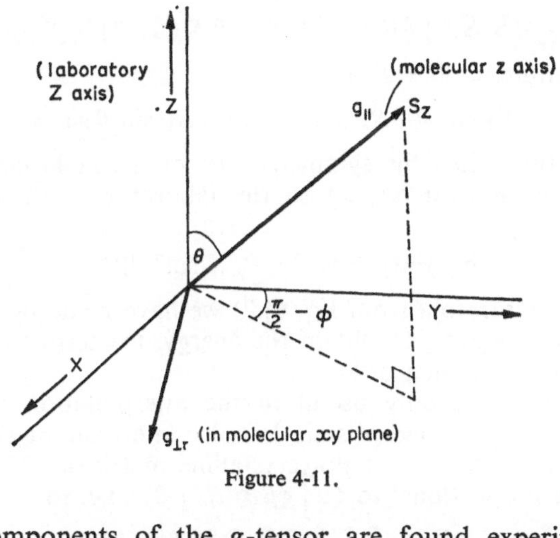

Figure 4-11.

The components of the g-tensor are found experimentally by mounting a single crystal in the cavity of an ESR spectrometer, then obtaining spectra at various points of rotation of the crystal, relative to the field, in three mutually perpendicular planes. Usually it is more convenient to rotate the magnet assembly than the cavity containing the crystal, and it is simplest to carry out the rotations about each axis separately. One way of achieving this is to remount the crystal at right-angles to its orientation on a previous series of measurements, implying that it is then only necessary for the heavy magnet to be rotated about a single axis. The crystal is set up using a goniometer.

As an example we shall consider an axially-symmetric radical. The g-value parallel to the axis of symmetry is $g_{\parallel}$ and that perpendicular to this axis, $g_{\perp}$. The "natural" system of coordinates for the molecule is xyz, the z-axis being parallel to the symmetry axis, and the laboratory coordinates, which we fix, are to be XYZ; the field is applied along the Z-axis.

The Hamiltonian for the system is greatly simplified by the element of symmetry and it is convenient to write it in terms of the molecular coordinates, first; i.e.:

$$\mathscr{H}=g_{\parallel}\beta B_z S_z+g_{\perp}\beta(B_x S_x+B_y S_y) \qquad (4.6)$$

This can be rewritten in the following form:

$$\mathscr{H}/\beta=g_{\perp}\bar{B}.\bar{S}+(g_{\parallel}-g_{\perp})\,B_z S_z$$

if now we apply this Hamiltonian twice, and make use of the anti-commutation properties of the components of the spin, which follow from the results of 2.6, i.e. $S_x S_y+S_y S_x=0$, we obtain,

$$\mathscr{H}^2/\beta^2=g_{\perp}^2(\bar{B}.\bar{S})^2+\tfrac{1}{2}[(\bar{B}.\bar{S})\,B_z S_z+B_z S_z(\bar{B}.\bar{S})]\,(g_{\parallel}^2-g_{\perp}^2) \qquad (4.7)$$

now we substitute for $B_z=B_Z\cos\theta$,

$$S_z=S_Z\cos\theta+S_X\sin\theta\cos\phi+S_Y\sin\theta\sin\phi$$

and we find that when the symmetry axis of the molecule is inclined at an angle of θ with respect to the laboratory axes, the g-value observed is

$$g(\theta)=(g_{\parallel}^2\cos^2\theta+g_{\perp}^2\sin^2\theta)^{1/2} \qquad (4.8)$$

In deriving this formula from Eq. (4.7) we have made use of the fact that when we integrate, to obtain the energy, the terms in S_x and S_y go to zero, i.e. to the first order.

Eq. (4.8) is not only useful in the interpretation of magnetic resonance spectra of single crystals, but also in evaluating the g-tensors of radicals in amorphous or polycrystalline matrices. The transition probability is proportional to $\langle\alpha\mid g\beta B_x' S_x\mid\beta\rangle^2$, i.e. to

$$gB_x'\langle\alpha\mid S_+ + S_-\mid\beta\rangle^2=(gB_x')^2 \qquad (4.9)$$

(B_x' is the microwave field).

Hence when all of the radicals are similarly orientated, the intensity of absorption will be approximately proportional to $[g(\theta+\pi/2)]^2$. This is the case with single crystal studies.

In polycrystalline material, however, the radicals will be pointing in all directions, at random, and at any particular field and frequency, the absorption of energy will arise only from those at an angle θ to the field, satisfying the resonance condition:

$$h\nu = g(\theta)\,\beta B$$

Now the number of radicals inclined so that their axes lie within a given solid angle, $d\omega$, will be proportional to that solid angle, so the fraction whose axes are between θ and $\theta + d\theta$ to the field, will be:

$$d\omega/4\pi = \tfrac{1}{2}\sin\theta\, d\theta = \tfrac{1}{2}\, d(\cos\theta) \tag{4.10}$$

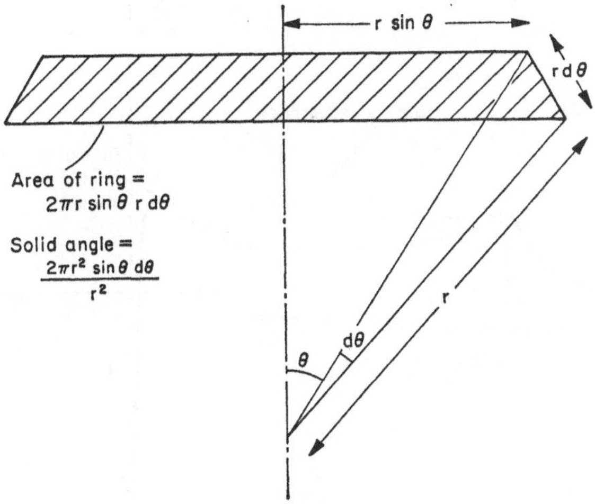

Figure 4-12.

We can now obtain an expression for the number of radicals absorbing between fields $\bar{B}$ and $\bar{B} + d\bar{B}$, i.e.

$$\tfrac{1}{2}\{(d/dB)\cos\theta\}\, dB$$

i.e.

$$dN/dB = -\tfrac{1}{2}g^2/B(g_{\parallel}^2 - g_{\perp}^2)\cos\theta$$

where we have used Eqs. (4.8) and (4.10).

The intensity of absorption at any point will be proportional to dN/dB and also to the transition probability, which also depends on the g-value, i.e. it is proportional to $g^2(\theta + \pi/2)$ according to Eq. (4.9) with $\bar{F} = g_{xx}\beta B_x' S_x$, B_x' being the magnetic component of the resonant radiation. The line shape will be given by the function $g^2(\theta + \pi/2)\, dN/dB$, and is shown with the smoothed out derivative spectrum in Figure 4-13.

Anisotropy in Hyperfine Splittings

In the solid state the coupling between spins is not an average as in solution, but will depend on the relative orientations of the elementary magnets. When the hyperfine interaction is largely determined by the classical, dipolar terms in the Hamiltonian, the magnetic resonance spectrum of the solid will change markedly with the orientation of the free radicals in the applied field. This is the case, for example, with the α-coupling constants in aliphatic radicals.

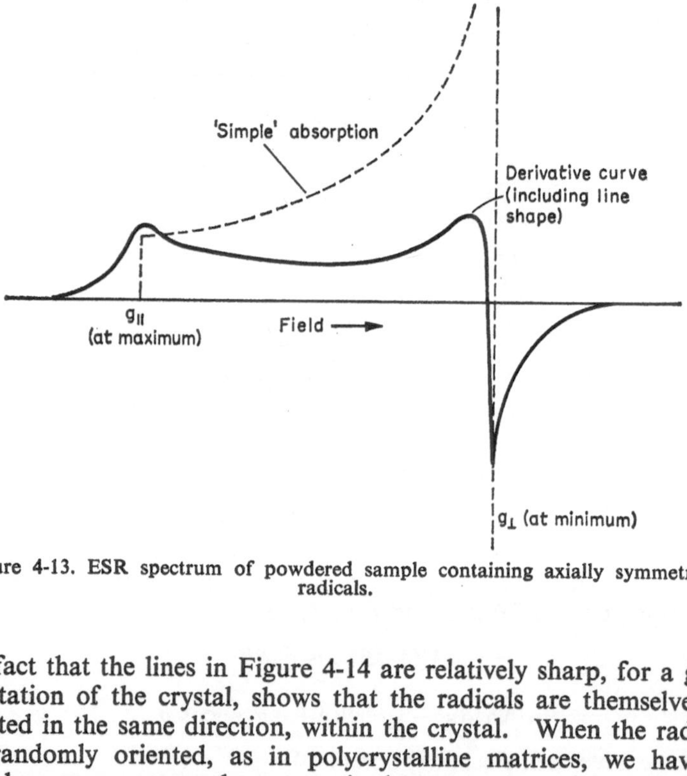

Figure 4-13. ESR spectrum of powdered sample containing axially symmetrical radicals.

The fact that the lines in Figure 4-14 are relatively sharp, for a given orientation of the crystal, shows that the radicals are themselves all oriented in the same direction, within the crystal. When the radicals are randomly oriented, as in polycrystalline matrices, we have to consider a space-averaged spectrum in the same way as that above for a species with an axially symmetric *g*-tensor. The situation in the presence of hyperfine coupling is doubly complicated by the fact that the principle values of the *g* and hyperfine coupling tensors are not necessarily in the same directions. As a simple example, however, we shall estimate what sort of spectrum we would expect from an odd electron interacting with a proton in such a way that both of these

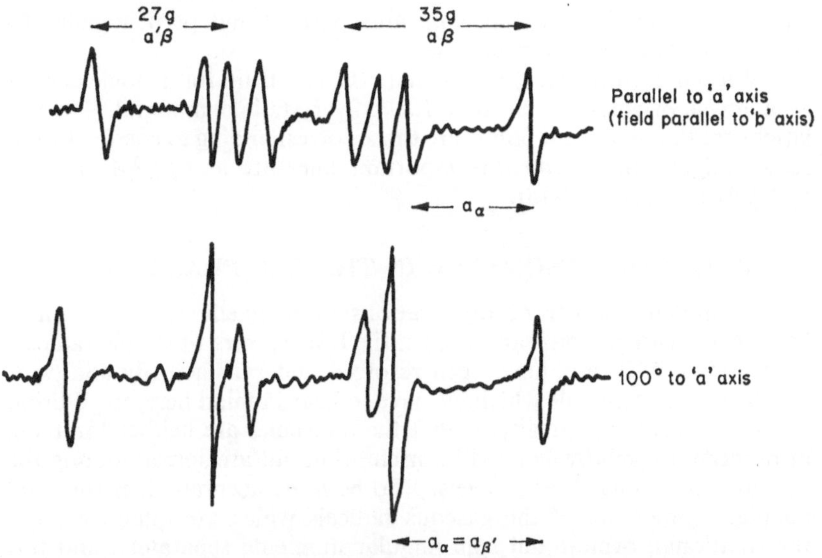

Figure 4-14. ESR spectra from single crystal of β succinic acid irradiated by x-rays (copied by kind permission of Taylor and Francis Ltd)

tensors are axially symmetric about the same axis. In this case we can write the Hamiltonian as:

$$\mathscr{H} = g(\theta)\, \beta B.S + T(\theta)\, S.I \tag{4.11}$$

where $g(\theta)$ is the same as before and the hyperfine interaction

$$T(\theta) = (A_\parallel^2 g_\parallel^2 \cos^2\theta + A_\perp^2 g_\perp^2 \sin^2\theta)^{1/2}/g(\theta)$$

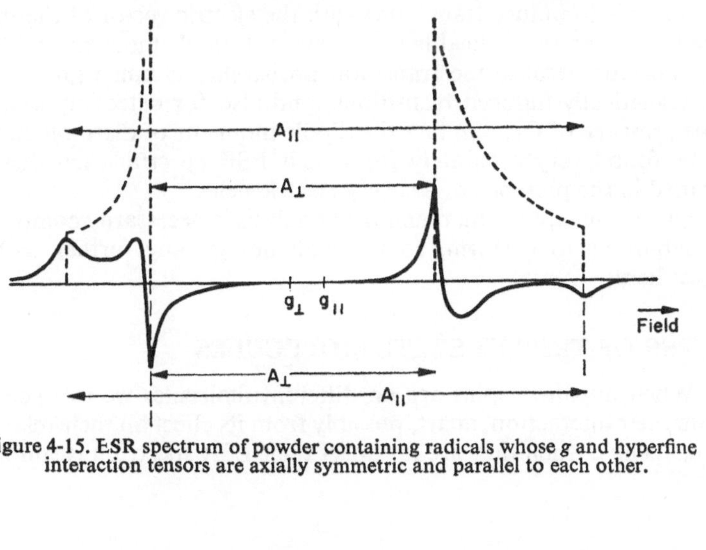

Figure 4-15. ESR spectrum of powder containing radicals whose g and hyperfine interaction tensors are axially symmetric and parallel to each other.

$A_{\parallel}$ and $A_{\perp}$ are the coupling constants parallel and perpendicular to the axis of symmetry, respectively.

We can use Eq. (4.11) to calculate the spectrum for a single crystal at a given orientation, or that for polycrystalline material. In the latter case the significant points are those corresponding to cos $\theta = 1$ or 0, as we might expect, and the important lines are at $g_{\parallel} \pm \frac{1}{2}A_{\parallel}$ and at $g_{\perp} \pm \frac{1}{2}A_{\perp}$ (see Figure 4-15).

4.5. MAGNETIC RESONANCE IN THE GAS PHASE

A limited number of observations of magnetic resonance have been made with gaseous samples, i.e. the ESR spectra of simple radicals such as SO, ClO, etc. have been resolved and satisfactorily analysed.

The main principle which we have to bear in mind here, in addition to those considered already, is that the molecules are neither tumbling at random nor relatively fixed in orientation but are spread among the available rotational energy levels. To be more accurate, it is the total angular momentum of the gaseous radicals which are quantized, not the rotational, orbital and spin angular momenta separately, and this means that the "spin" resonance will be affected not only by the orbital angular momentum, but also by the rotations of the molecule as a whole.

The magnetic energy levels depend then on:

(a) The rotational level (lowest is most important).
(b) The usual Zeeman interaction (including spin–orbit coupling).
(c) Hyperfine interactions with the nuclei, i.e. scalar (Fermi contact interaction), dipolar and quadrupolar coupling.

It is possible to induce transitions with the electric vector of the microwave field when the radical has a permanent dipole moment and this is advantageous because the transition probability is much greater than for magnetically induced transitions, and also for detecting, say, ClO in the presence of O_2. In fact the dipole moments of diatomic radicals can be found very accurately from their ESR spectra when these are obtained in the presence of a steady electric field.

Unfortunately the mathematical analysis is necessarily complicated and laborious to perform, so we shall not go any further with the subject here.

4.6. ESR OF TRIPLET STATE MOLECULES

When unpaired spins are on different molecules we can generally ignore their interaction, apart, possibly from its effect on their relaxation times. Thus, when they are far apart their ESR spectrum is simply the sum of their individual spectra. As the spins are held closer and closer together, e.g. by being in the same molecule, their overall spectrum becomes inevitably modified by their dipole–dipole interaction. The appropriate anisotropic term in the Hamiltonian is given by Eq. (1.8), i.e.

$$\mathscr{H}=g\beta\bar{B}(\bar{S}_1+\bar{S}_2)+g^2\beta^2[\bar{S}_1.\bar{S}_2/r^3-3(\bar{S}_1.r)(\bar{S}_2.r)/r^5] \quad (4.12)$$

where $\bar{S}_1$, $\bar{S}_2$ are the spins of electrons 1, 2 respectively, and $\bar{r}$ the vector distance between them.

Eq. (4.12) is not of much use to us as it stands, because we can measure the effects only of S_1+S_2 and not of S_1 or S_2 separately. However, if we can assume that the eigenfunctions of this Hamiltonian resemble products of the eigenfunctions of S_1 and S_2, i.e. $\alpha_1\alpha_2$, $\beta_1\beta_2$, $(\alpha_1\beta_2+\beta_1\alpha_2)/\sqrt{2}$ then we can reduce it to an expression containing the total spin only:

$$\mathscr{H}_{\text{aniso}}=\bar{S}.\bar{\bar{D}}.\bar{S} \quad (4.13)$$

where $\bar{\bar{D}}$ is a tensor whose nine elements depend on the spatial co-ordinates, or rather their mean values, e.g.

$$D_{xx}=\tfrac{1}{2}g^2\beta^2\left\langle\frac{r^2-x^2}{r^5}\right\rangle$$

$\bar{S}$ is the total spin $\bar{S}_1+\bar{S}_2$.

In deriving Eq. (4.13) extensive use is made of the commutation properties of the components of the two spins, i.e. the components of $\bar{S}_1$, say, commute with all the variables except the other components of $\bar{S}_1$.

It is customary to write Eq. (4.13) in the coordinate system in which $\bar{\bar{D}}$ is diagonal, i.e.

$$\mathscr{H}_{\text{aniso}}=-XS_x^2-YS_y^2-ZS_z^2 \quad (4.14)$$

Since the trace of $\bar{\bar{D}}$ is zero, i.e. $D_{xx}+D_{yy}+D_{zz}=0$, X, Y and Z are not independent, for also $X+Y+Z=0$, and we can, if we wish, write the Hamiltonian as

$$\mathscr{H}=g\beta H.S+D(S_z^2-S^2/3)+E(S_x^2-S_y^2) \quad (4.15)$$

The easiest way of obtaining the eigenvalues of this Hamiltonian is to start with spin functions which correspond with the diagonal form of the anisotropic part given in Eq. (4.14). These are $(|\beta\beta\rangle-|\alpha\alpha\rangle)/\sqrt{2}$, $i(|\alpha\alpha\rangle+|\beta\beta\rangle)/\sqrt{2}$ and $(\alpha\beta+\beta\alpha)/\sqrt{2}$. In terms of these basis functions the Zeeman interaction is antisymmetric, i.e.

its diagonal elements are zero. When the applied field is in the z-direction we have to solve the secular determinant

$$\begin{vmatrix} X-E & -ig\beta H & 0 \\ ig\beta H & Y-E & 0 \\ 0 & 0 & Z-E \end{vmatrix} = 0 \tag{4.16}$$

The energy levels are then $E=Z,\ -\frac{1}{2}Z\pm\sqrt{(X-Y)^2/4+g^2\beta^2H^2}$. It is now relatively easy to plot the change in energy as the applied field changes, for example, when the applied field is zero $E=X$, Y or Z. For a given microwave frequency there are two transitions corresponding to $\Delta m_s=1$, and another, weaker transition corresponding to $\Delta m_s=2$, i.e. a "double" jump which is a second-order effect

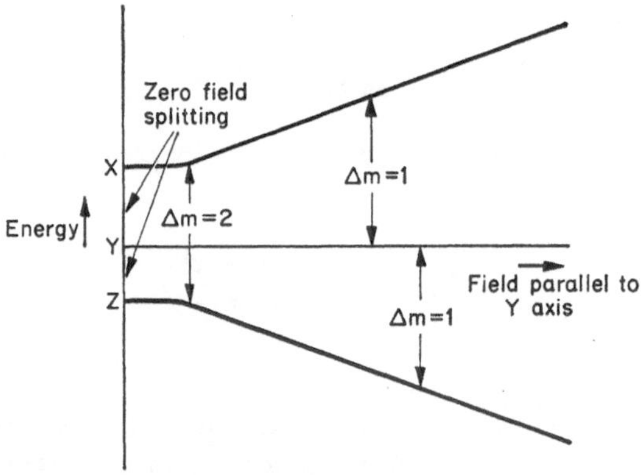

Figure 4-16. Energy level diagram illustrating zero field splitting of levels in a triplet state molecule.

Triplet state molecules are generally very difficult to observe in solution because of the unfavourable relaxation times, which lead to broadening of the signals. It is also difficult to find crystals in which such molecules may be dispersed and constantly oriented, favourable examples being naphthalene or phenanthrene in single crystals of durene.

Most studies of triplet states by ESR are therefore carried out in frozen glasses, and in order to interprete them we have to consider, as in the previous section, an average of the spectra of molecules having random but fixed orientations. The derivation of parameters may be greatly facilitated if the $\Delta m_s=2$ transition can be satisfactorily observed, since it is relatively isotropic and will therefore have rather sharper "averaged" lines than the more intense, but anisotropic, $\Delta m_s=1$ transitions.

As in calculating the spectra of doublet species in polycrystalline materials, we can expect sudden changes in absorption when the applied field is parallel to the principle axes; in the present case these are of the dipole–dipole interaction tensor. Because of the zero-field splitting there are six "steps" as shown in Figure 4-17.

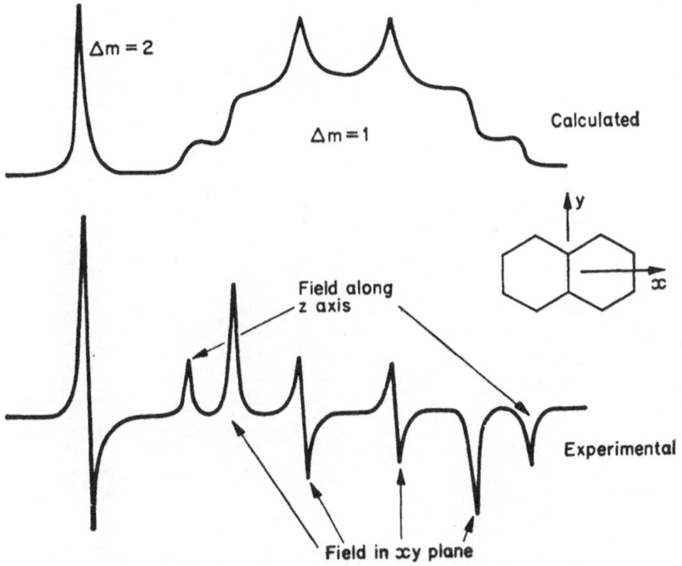

Figure 4-17. ESR spectrum of naphthalene in its triplet state.

It is of interest to note that the $\Delta m_S = 2$ transition occurs at half-field, since the one quantum of radiation has effectively to turn over two spins at once.

Chapter 5

Theory of Chemical Shifts and *g*-values. The Influence of Electronic Orbital Angular Momentum on the Position of Resonance

In the previous chapter we discussed some ways in which the magnetic resonance spectrum of a particle can be modified by an environment consisting of other charged and/or magnetic species. Empirically we determine certain parameters, which are generally tensor quantities, having nine components, since they (the particles) are in three dimensional space, and are termed *g*-values, chemical shifts and coupling constants. The numbers actually obtained for these "time-independent" parameters depend, in the first instance, on the composition and electronic structure of the atom, molecule or group of molecules in which the nucleus or odd electron under investigation is embedded.

It is convenient to classify the magnetic interactions of the particle whose resonance is being observed, into those with other spins, which lead to coupling constants, Knight shifts etc., and those which arise from the relative motion of other charged particles in the system. It is the latter type of interaction which we shall consider in this chapter, because this is what determines the field-dependent position of the resonance spectrum compared with that of the isolated particle.

5.1. EXPRESSIONS FOR THE MAGNETIC INTERACTIONS

When a magnetic field is applied to a particle in a molecule, changes in the electronic motion take place which can be described in various ways. We shall adopt the most common approach which is to work in a gauge in which the vector potential of the field is given by:

$$\bar{A} = \tfrac{1}{2}\bar{B} \wedge \bar{r}$$

In this case, application of a field leads to a first-order precession of the electrons about the direction of that field. According to Lenz's law, the direction of precession is such as to oppose the applied field, so this is a diamagnetic effect. A nucleus of an atom therefore experiences a smaller field than that applied, due to this "Larmor" precession of electrons. The magnitude of this diamagnetic effect is, however, not observable, since it depends on our choice of gauge. Only when we take the calculation to the second order do we obtain an expression which is gauge-invariant and therefore, observable. The second-order term describes an effect opposite to that of the first-order one, i.e. a paramagnetic effect, and it depends on the way the motion of the electrons is modified by the presence of spins and also by the applied field (see figure 5-1).

Another way of looking at the situation is that there is a three-way coupling, between the applied field, the orbital motion of the electron or electrons, and the spin of the particle concerned, and that the interplay of these is what determines the position of the centre of its magnetic resonance spectrum. It is helpful to consider a diagrammatic summary of the principles involved:

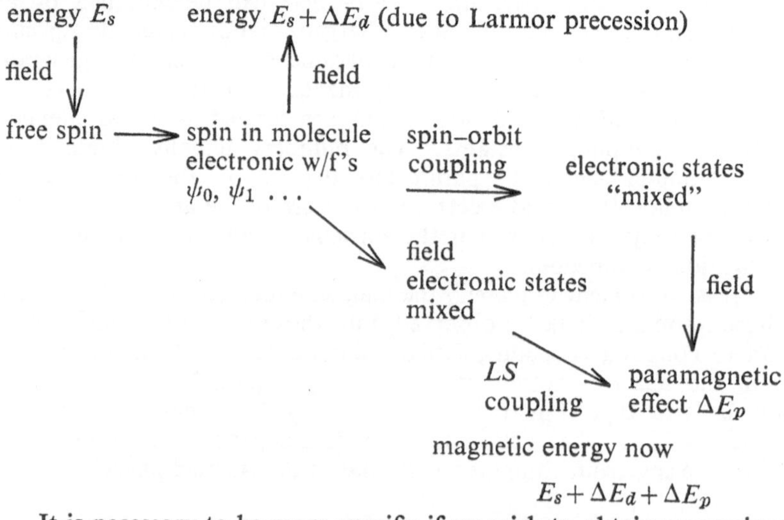

It is necessary to be more specific if we wish to obtain expressions for the magnetic energy levels of a particle in a field, so we write down again the appropriate terms in the Hamiltonian. For simplicity we shall write formulae involving only one nucleus and one electron.

(i) *Electronic Zeeman Interaction*, i.e. the direct interaction between the electronic moment (combined orbital and spin parts) and the applied field.

$$-(\bar{\mu}_L+\bar{\mu}_S).\bar{B}=\beta_e(\bar{L}+g_e\bar{S}).\bar{B}=\beta_e\bar{J}.\bar{\bar{g}}.\bar{B}$$

where $\bar{L}\hbar$, $\bar{S}\hbar$ and $\bar{J}\hbar$ are the orbital, spin and total angular momenta respectively: $g_e=2$; $\beta_e=(e/2mc)\hbar=$ Bohr magneton.

(ii) *Nuclear Zeeman Interaction*, i.e. the direct interaction between the nuclear moment and the applied field

$$-\bar{\mu}_N.\bar{B}=-\gamma_N\hbar\bar{B}.\bar{I}=-\beta_N g_N\bar{B}.\bar{I}$$

where $I=$ nuclear spin, γ_N, the magnetogyric ratio

$$\beta_N=(e\hbar/2M_Nc)=\text{nuclear magneton}$$

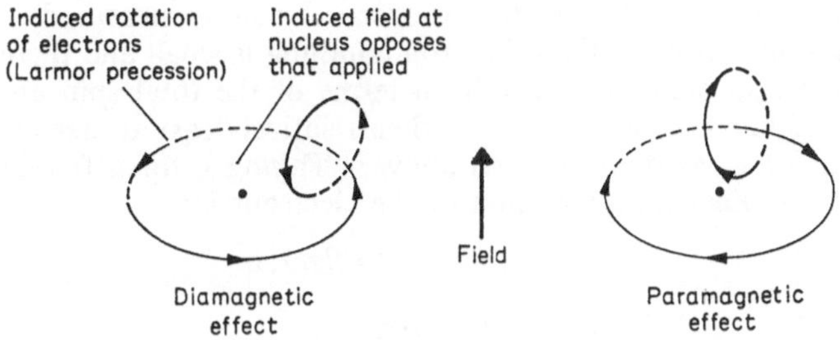

Figure 5-1.

(iii) *Electron Spin–Orbit Interaction*, i.e. the electromagnetic coupling between the electron spin and the nuclear charge, due to their relative motion.

$$-\bar{\mu}_e.\bar{E}\wedge\bar{v}/2c=\zeta_e\bar{L}.\bar{S}$$

where $\bar{E}=$ electric field at the electron due to the nuclear charge

$$\zeta_e=Z(e^2\hbar^2/2m^2c^2)\langle 1/r^3\rangle=\text{spin–orbit coupling constant}$$

(iv) *Nuclear Spin–Electron Orbit Interaction*, i.e. the electromagnetic coupling between the nuclear spin and the electronic charge due to their relative motion.

$$-\bar{\mu}_N.\bar{E}_\wedge\bar{v}/c=\zeta_N\bar{L}.\bar{I}$$

where $-\bar{E}=$ electric field at the nucleus due to the electronic charge

$$\zeta_N=(\gamma_N\hbar^2/mc)\langle e/r^3\rangle$$

In (iii) and (iv) we have written the interactions in the same form to show the symmetry, in spite of the fact that in the case of the electronic spin–orbit coupling, the spin and the orbital motion are associated with the same particle, relative to an observer at rest relative to the nucleus.

As we shall see, the role of the electronic spin–orbit coupling in modifying the g-value of an electron is the same as that of the nuclear spin and the electronic charge in determining the chemical shift of the nuclear moment. However, the importance of different terms varies according to the situation, for example, in proton resonance, the local electronic orbital angular momentum is zero, at least in the first instance, so the "small" term arising from the Larmor precession of the electrons is unusually important.

5.2. g-VALUES AND CHEMICAL SHIFTS IN ATOMS

Isolated atoms are especially simple because their electron distributions are spherically symmetrical. One of the implications of this symmetry is that the numerical magnitude and one component of the total angular momentum of the electrons, J, can be measured. In the Russell-Saunders case, the spin–orbit coupling is small and therefore it is a good approximation to talk in terms of the total spin and total orbital angular momentum. It is then relatively easy to derive an expression for g, as defined in (i) above. Taking g for a free electron to be 2, the Zeeman interaction of the electrons is:

$$\beta(L+2S).B=\beta g_e J.B$$

i.e.

$$(L+S)(L+2S)=gJ^2$$

i.e.

$$J^2+S^2+L.S=gJ^2$$

but

$$2L.S=J^2-L^2-S^2$$

therefore

$$\tfrac{1}{2}(3J^2-L^2+S^2)=gJ^2$$

replacing the operators by their eigenvalues, e.g. $J^2\,|\,\rangle=\hbar^2 j(j+1)\,|\,\rangle$ etc., we obtain the Lande spectroscopic splitting factor:

$$g=3/2-[l(l+1)-s(s+1)]/2j(j+1) \tag{5.1}$$

One might expect that for the lighter elements (small nuclear charge) measurements involving ESR, bulk magnetic susceptibilities, or atomic spectra, should lead to values of g consistent with Eq. (5.1), allowing for suitable modifications due to relatively small spin–orbit interactions. Such is the case with gaseous atoms and with the ions of the rare earths. Solutions of paramagnetic ions of the first and second transition series, however, behave magnetically as if the unpaired electrons had virtually no orbital angular momentum, i.e. as if they had spin angular momentum only. Similarly, the g-values of most free radicals are very close to the free spin value, which implies, as a matter of interest, that their ESR spectra generally overlap, making it difficult to examine more than one radical at a time using this method. An analogous situation arises in NMR spectra, that is, the first-order interaction between the nuclear spin and any electronic orbital angular momentum is zero, in the condensed phases or in molecules.

5.3. QUENCHING OF ORBITAL ANGULAR MOMENTUM

The apparent loss of orbital angular momentum in going from an atomic to a molecular system, i.e. on introducing a relatively strongly interacting environment, is due to the loss of symmetry and is called the quenching of orbital angular momentum by a ligand field. It is most easily examined in terms of crystal field theory.

Let us consider as a simple example, a free atom with one electron in the degenerate p orbitals. The eigenfunctions of the energy may also be eigenfunctions of L^2 and of L_z, so we may write them as $|+1\rangle$, $|-1\rangle$ and $|0\rangle$, corresponding to $l=1$, $m=\pm 1$ and 0.

Observed values of L_z will therefore be ± 1 or zero.

In view of the degeneracy of the levels, we can write the linear combinations

$$\psi(p_z) = |0\rangle,$$

$$\psi(p_x) = \frac{1}{\sqrt{2}}(|+1\rangle + -1\rangle),$$

$$\psi(p_y) = \frac{i}{\sqrt{2}}(|+1\rangle - |-1\rangle)$$

and still have eigenfunctions of the Hamiltonian. These are the familiar real forms of the orbitals and they are also eigenfunctions of L^2 but not of L_z.

Now let us suppose that the atom is placed in an electric field, say due to ligands placed octahedrally around it, which removes the degeneracy, so that the p_x, p_y and p_z orbitals have different energies E_x, E_y and E_z respectively (see figure 5-2).

We can measure the expectation values of the components of the angular momentum, one at a time, by means of the interaction with a magnetic field applied consecutively along appropriate directions. Leaving out the effects of spin for the moment, a field applied along a direction chosen as the z axis, leads to a perturbation $\beta L_z H_z$, i.e. the Hamiltonian changes from H_0 to $H_0 + \beta L_z H_z$. In order to obtain good approximations to the new energy levels, we can start from the eigenfunctions prior to the application of the field. The new orbitals may be written approximately as linear combinations of these,

$c_1\,\psi(p_z)+c_2\,\psi(p_x)+c_3\,\psi(p_y)$ and we can use the variational principle to find the modified energies. The secular determinant is:

$$\begin{vmatrix} E_z-E & 0 & 0 \\ 0 & E_x-E & -i\beta B_z \\ 0 & +i\beta B_z & E_y-E \end{vmatrix}=0$$

and the solutions are:

$$E=E_z, \qquad (E_x+E_y)/2\pm\tfrac{1}{2}\sqrt{(E_x-E_y)^2+\beta^2 B_z{}^2} \tag{5.2}$$

From this result we see that if the ligand field splitting, E_x-E_y, is large compared with βB_z, which is usually the case, the energies will not change appreciably when a magnetic field is applied, i.e. $\langle L_z\rangle\approx 0$, and the orbital angular momentum is said to be quenched.

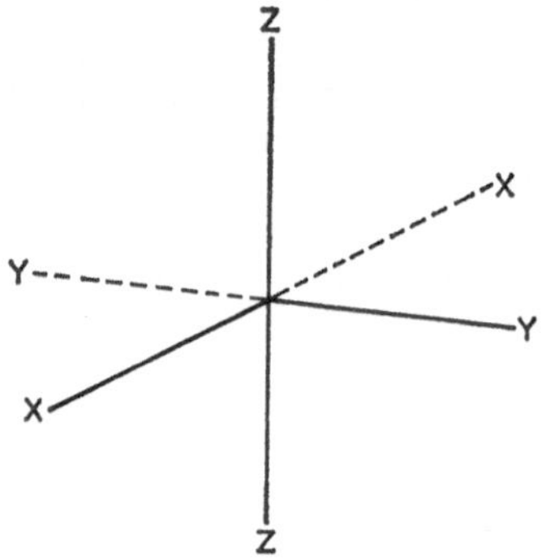

Figure 5-2.

What is happening here is that the p_x, p_y p_z orbitals are eigenfunctions of the Hamiltonian with or without the ligand field. On the other hand the eigenfunctions of L_z, $|+1\rangle$, $|-1\rangle$ are not eigenfunctions of the Hamiltonian in the presence of the ligand field. The expectation value of L_z is zero because the p_x, p_y and p_z orbitals are all associated with zero angular momentum about any particular axis, e.g.

$$\langle p_x\,|\,L_z\,|\,p_x\rangle=\tfrac{1}{2}\langle +1\,|+\langle -1\,|\,)\,L_z(\,|+1\rangle+|-1\rangle=0.$$

On the other hand, when the ligand/crystal field is weak, as it is with relatively free atoms or with the rare earth ions, when the odd 4f electrons are effectively shielded from the ligand field, the energy does depend on the applied magnetic field, i.e.

$$E\approx E_z, \quad (E_x+E_y)/2\pm\beta H_z$$

In these cases, then, the orbital angular momentum is not quenched, the eigenfunctions are also those of L_z, i.e. $|+1\rangle$, $|-1\rangle$ and $|0\rangle$, and the g-values are given by a formula such as that in Eq. (5.1).

We can look at all of this from a slightly different point of view, and at the same time, generalize a little: suppose an eigenfunction is $a+ib$, then the expectation value of the operator

$$L_z = \frac{\hbar}{i}\frac{\partial}{\partial\phi}$$

is given by the integral:

$$\langle L_z\rangle = -i\hbar \int_0^{2\pi} (a-ib)\frac{\partial}{\partial\phi}(a+ib)\,d\phi \qquad (a, b \text{ are real})$$

Here ϕ is the angle (of rotation about the z axis), and since the probability function, a^2+b^2 must be single-valued, it must have the same value at $\phi=2\pi$ as it has at $\phi=0$. The above equation becomes:

$$\langle L_z\rangle = \hbar \int_0^{2\pi} \left(a\frac{\partial b}{\partial\phi} - b\frac{\partial a}{\partial\phi}\right) d\phi$$

So in the general case, $\langle L_z\rangle$ will be real and non-zero. When the eigenfunctions are either real ($b=0$) or purely imaginary ($a=0$), only zero values will be found for L_z, i.e. the orbital angular momentum will be quenched.

5.4. THE EFFECTS OF SPIN–ORBIT COUPLING

In view of these results, we might wonder why all g-values are not the free spin value, and why chemical shifts are not always solely due to the Larmor precession of electrons. The reason why orbital angular momentum does have a marked influence on these parameters, in many cases, lies in the spin–orbit terms given in (iii) and (iv) in our list in **5.1**.

In the first place, as soon as an external field is applied, the electron orbits all start precessing about the nucleus, which is taken to be stationary relative to the field, according to the Larmor theorem. Nuclear spins therefore experience additional forces due to this precessional motion of the electrons. Similarly an electron spin is also subject to extra forces due to the relative precession of the nuclei, but these forces are generally considered negligible. Using the more appropriate example, the first-order correction to the Zeeman energy of a nucleus, due to its interaction with the electronic motion is, from

(iv) in **5.1**:

$$-\bar{\mu}_N.\bar{\sigma}.\bar{B} = -\langle 0 \mid \bar{\mu}_N.\bar{E} \wedge \bar{v} \mid 0\rangle/c$$
$$= -\langle 0 \mid \bar{\mu}_N L + e\bar{\mu}_N.\bar{E} \wedge (\bar{B} \wedge \bar{r})/2c \mid 0\rangle$$
$$= +e^2\bar{\mu}_N\langle \bar{B}/r - \bar{r}(\bar{B}.\bar{r})/r^3\rangle/2mc^2$$

using the results of the previous section and a little vector algebra. We can now write expressions for the components of the chemical shift tensor, e.g.

$$\mu_z\sigma_{zz}B_z = +e^2\mu_z\langle B_z/r - z(B_z z)/r^3\rangle/2mc^2$$

so that

$$\sigma_{zz} = +(e^2/2mc^2)\langle (x^2+y^2)/r^3\rangle$$

In solution we are usually more interested in the spatial average of the chemical shift which is given by $(\sigma_{xx}+\sigma_{yy}+\sigma_{zz})/3$; so the first-order contribution to the chemical shift is given by:

$$\sigma_d = (e^2/3mc^2)\langle 1/r\rangle \tag{5.3}$$

Eq. (5.3) is equivalent to the Lamb formula for diamagnetism, and it tells us the susceptibility of the medium at the nucleus.

The Second-Order Perturbation Term

Now we come to what is generally the more important contribution to *g*- or chemical shifts, and that is the paramagnetic term, as it is called in NMR, and the only one which is usually considered in ESR. It is worthwhile writing down the formula for the second-order perturbation energy again, i.e.

$$E_2 = \Sigma_n\langle 0 \mid H' \mid n\rangle\langle n \mid H' \mid 0\rangle/(E_0 - E_n)$$

where H' is the perturbation and "n" a label for the nth excited state. The terms in this expansion which are of interest to us in the present context are the cross-terms between the Zeeman and the spin–orbit interactions. For an electron spin these are:

$$E_2 = \Sigma_n 2\langle 0 \mid \zeta_e L.S \mid n\rangle\langle n \mid \beta L.B \mid 0\rangle/(E_0 - E_n) \tag{5.4}$$

We can rewrite Eq. (5.4) in tensor form and also expand it a little to avoid errors later:

$$E_2 = -\beta\bar{S}.[\Sigma_n(\langle 0 \mid \zeta_e\bar{L} \mid n\rangle\langle n \mid \bar{L} \mid 0\rangle + \langle 0 \mid \bar{L} \mid n\rangle\langle n \mid \zeta_e\bar{L} \mid 0\rangle)/\Delta E_n].\bar{B} \tag{5.5}$$

where $\Delta E_n = -(E_0 - E_n) > 0$.

When the basis functions are eigenfunctions of L_z, or failing that, $|\,0\rangle$ is, then integrals such as $\langle 0\,|\,L_z\,|\,n\rangle$ will be zero. In this case the diagonal component g_{zz} of the g-tensor will be the same as the free spin value g_e, which is simply that obtained from first-order perturbation theory when the orbital angular momentum is quenched. We can see this from the following examples of the second-order contributions to components of the g-tensor:

$$\Delta g_{zz} = -2\Sigma_n\langle 0\,|\,\zeta_e L_z\,|\,n\rangle\langle n\,|\,L_z\,|\,0\rangle/\Delta E_n \;(=0)$$
$$\Delta g_{yy} = -2\Sigma_n\langle 0\,|\,\zeta_e L_y\,|\,n\rangle\langle n\,|\,L_y\,|\,0\rangle/\Delta E_n \qquad (5.6)$$

The corresponding first-order contributions are found from the energy,

$$\beta\langle 0\,|\,\bar{L}+g_e\bar{S}\,|\,0\rangle.\bar{B}=\beta g_e\langle\bar{S}\rangle.\bar{B}$$

i.e.

$$g_{zz}=g_e$$
$$g_{yy}=g_e+\Delta g_{yy}\ \ldots \text{ and so on.}$$

Simple example: Consider the situation when an odd electron is in the $2p_z$ orbital of an atom placed in such a field that the $2p_x$ and $2p_y$ orbitals are degenerate but higher in energy than the $2p_z$ orbital. Take $g_e=2$, and since the $2p_x$ and $2p_y$ orbitals are degenerate, it is easier to use $|+1\rangle$ and $|-1\rangle$ as our basis functions.

As in the above case, $|\,0\rangle$ is an eigenfunction of L_z so $\Delta g_{zz}=0$. We can approach the infinite series given in Eq. (5.6) in several ways in the first we might neglect higher excited states, in which case we find:

$$\Delta g_{xx} = -2(\langle 0\,|\,\zeta_e L_+\,|-1\rangle\langle -1\,|\,L_-\,|\,0\rangle + \langle 0\,|\,\zeta_e L_-\,|+1\rangle\langle +1\,|\,L_+\,|\,0\rangle)/4\Delta E$$

i.e.

$$\Delta g_{xx} = -2\zeta_e/\Delta E$$

where ΔE is the lowest excitation energy and we have replaced L_x by $\frac{1}{2}(L_+ + L_-)$, using the relations such as

$$L_+\,|-1\rangle=\sqrt{2}\,|\,0\rangle, \qquad L_-\,|+1\rangle=\sqrt{2}\,|\,0\rangle,$$

and so on (see Appendix).

The resultant expressions for the diagonal elements of the g-tensor are given by the equations:

$$g_{zz}=2$$
$$g_{xx}=g_{yy}=2-2\zeta_e/\Delta E \qquad (5.7)$$

Chemical Shift—The Paramagnetic Contributions

We can obtain the second-order contribution to the chemical shift of the nucleus in the above case directly from the *g*-shift. This is possible because to find the change in g_N we simply start by considering $\zeta_N \bar{L}.\bar{I}$, instead of $\zeta_e \bar{L}.\bar{S}$, in the Hamiltonian, so the relevant terms are:

$$H'_{(L.I.)} = -\gamma_N \hbar \bar{B}.\bar{I} + \zeta_N \bar{L}.\bar{I} + \beta \bar{L}.\bar{B}$$

From the second-order terms which depend on the first power of *B* we obtain:

$$(\Delta g_N)_{zz} = 0, \qquad (\Delta g_N)_{xx} = (\Delta g_N)_{yy} = -(2\zeta_N/\Delta E)\frac{\beta_e}{\beta_N}$$

Translating these into chemical shifts, and taking an average, so that the result applies to solutions, we find:

$$\sigma_p = -(2e^2\hbar^2/3m^2c^2\Delta E)\langle 1/r^3\rangle \qquad \text{(low-field)} \qquad (5.8)$$

Comparing Eqs. (5.3) with (5.8) we see that the two contributions are of opposite signs and that the diamagnetic effect is of longer range (depending on $\overline{1/r}$ rather than $\overline{1/r^3}$). This is why bulk susceptibilities of molecules containing no unpaired electrons are negative, i.e. most substances are diamagnetic.

In the multi-electron systems, we have to sum over all the electrons to find the total chemical shift, for there is one term, such as Eqs. (5.3) or (5.8) for each electron.

The general formula for the paramagnetic part of the chemical shift is derived from the equation corresponding to Eq. (5.4), but with $\zeta_N I$ instead of $\zeta_e S$.

Averaged Energy Approximation

Another approach to the problem posed by the infinite series in Eq. (5.4) is to replace all the ΔE_n's by a sort of average, ΔE say, which is generally to be determined semi-empirically. The effect of this is that we over-estimate instead of under-estimate the higher excited states. The accompanying simplification is great, for we can now use the closure theorem: $\Sigma_r \mid r\rangle\langle r \mid = 1$. The series is thereby reduced to a single term depending only on the ground-state wavefunction:

$$E_2 = 2\beta\bar{S}.\langle 0 \mid \zeta_e \bar{L}\bar{L} \mid 0\rangle.\bar{B}/\Delta E \qquad (5.9)$$

This leads to formulae for the diagonal components of the *g*-tensor which are the same as those given in Eq. (5.7), in the example chosen, because

$$L_z^2 \mid 0\rangle = 0, \qquad L_x^2 \mid 0\rangle = 1 \mid 0\rangle.$$

In spite of some early successes, mainly in the calculation of nuclear spin–spin couplings, the averaged energy approximation is being used less and less now, largely because of the lack of any rigorous justification, and the ill-defined nature of ΔE.

5.5. THE EFFECTIVE SPIN HAMILTONIAN

There are several possible ways of looking at the spin Hamiltonian, $\beta\bar{S}.\bar{\bar{g}}.\bar{B}$. One is that the spin experiences not the applied field but one suitably modified by the interaction of the orbital moment and the applied field. This was effectively the approach we used above when we introduced the perturbations due to the applied field at the same time as that due to the spin–orbit coupling. However, the interaction between the spin and the electronic orbital angular momentum, whether the spin be electronic or nuclear, existed before the introduction of any applied field. It is therefore perhaps more satisfying to consider the interaction of the field with the atom in its ground-state which includes the effects of the spin–orbit coupling.

In the absence of a field, the effect of spin–orbit coupling is to mix up states with different values of m_L, m_S. For example, if before introducing the perturbation the ground-state wavefunctions were $|0, +\frac{1}{2}\rangle$ and $|0, -\frac{1}{2}\rangle$, then they become, under the influence of the term $\zeta_e L.S = \zeta_e[L_zS_z + \frac{1}{2}(L_+S_- + L_-S_+)]$:

$$\begin{aligned} |+\rangle &= |0, +\tfrac{1}{2}\rangle - \zeta_e\,|+1, -\tfrac{1}{2}\rangle/\sqrt{2}\Delta E \\ |-\rangle &= |0, -\tfrac{1}{2}\rangle - \zeta_e\,|-1, +\tfrac{1}{2}\rangle/\sqrt{2}\Delta E \end{aligned} \tag{5.10}$$

where as before we neglect higher excited states, and we have used the usual formula for the first-order wavefunctions (see Eq. (2.15)). As with the original doublet, these two wavefunctions belong to degenerate states and constitute a "Kramers doublet". The degeneracy is destroyed, of source, when a field is applied. Now we introduce an operator S' which behaves with $|+\rangle$ and $|-\rangle$ in the same way that the true spin S operates on $|+\frac{1}{2}\rangle$ and $|-\frac{1}{2}\rangle$, e.g.

$$S_z'\,|+\rangle = +\tfrac{1}{2}\,|+\rangle, \qquad S_z'\,|-\rangle = -\tfrac{1}{2}\,|-\rangle, \text{ etc.}$$

S' is called the effective spin because it contains a small amount of orbital angular momentum character which makes the g-tensor anisotropic.

The effective spin Hamiltonian is $\beta\bar{S}'.\bar{\bar{g}}.\bar{B}$ and this has to give the same matrix elements as the true Zeeman interaction, $\beta(\bar{L}+g_e\bar{S}).\bar{B}$ so that we get the same value for the energy on applying the field. The components of the g-tensor calculated in this way are the same as those obtained already by straight application of second-order perturbation theory: e.g.

$$\langle + \mid g_{xx}S_x' \mid + \rangle = \langle + \mid L_x + g_e S_x \mid + \rangle = 0$$

$$\langle + \mid g_{xx}S_x' \mid - \rangle = \langle + \mid L_x + g_e S_x \mid - \rangle$$

$$= \tfrac{1}{2}\langle + \mid L_+ + L_- + g_e(S_+ + S_-) \mid - \rangle$$

i.e.

$$g_{xx} = (g_e - 2\zeta_e/\Delta E)$$

where on the right-hand side of the equation we have used the expression given in Eq. (5.10) for $\mid + \rangle$ and $\mid - \rangle$.

We could develop a theory of chemical shifts in an exactly similar way, i.e. by replacing $\bar{S}$ by $\bar{I}$, g_e by g_N, etc. and by defining an effective nuclear spin. The equations so obtained would have the same form as those involving electron spins.

5.6. PREDICTIONS OF *g*-FACTORS AND CHEMICAL SHIFTS

At the outset it is necessary to recognize that the use of simple valence theories in "explaining" features of magnetic resonance spectra, is only justifiable if we use them cautiously—almost suspiciously. The most we can hope for is that either they will lead to correct qualitative predictions about grosser effects, such as whether to expect a resonance to occur at high or low field in a particular case, or that they will enable us to correlate results obtained for closely related series of compounds.

Estimation of *g*-Values

In general, the components of *g*-tensors are estimated semi-empirically by finding the spin–orbit coupling constant ζ_e from atomic spectra and the excitation energy ΔE from the spectrum of the complex/molecule under discussion. Expressions for the *g*-shifts then depend on these parameters and the symmetry, or lack of symmetry of the complex, and are found using either crystal field theory or by some more complex procedure, e.g. ligand field theory. Each molecule tends to present a special case, and there seem to be few generalizations one can make. We shall only mention two cases where it seems that we can get at least a qualitative idea about the direction of *g*-shifts.

In the example given above, there was one electron in an otherwise empty shell of orbitals whose degeneracy is split by a ligand field. In this case the first excited state corresponds to an excitation of the odd electron to an orbital of higher energy (see Figure 5-3). This leads to an average *g*-value which is *less* than the free spin value. An example of this is the $3d^1$ system of Ti^{III} in $NaTi(SO_4)_2 . 12H_2O$, $\bar{g} = 1.2$.

On the other hand, in a situation which is similar, except that there is one "hole" in the electronic sub-shell, the 1st excited state involves the movement of the odd electron into a *lower* atomic orbital, and the ratio $\zeta_e/\Delta E$ of the odd electron will be negative. Hence, other things being equal, the *g*-value will be *greater* than the free electron value. An example of this is the d^9 system of Cu^{II} in $CuSO_4$, $5H_2O$, $g=2.2$. Another way of looking at this particular situation is that the electron which is excited has the opposite spin to that of the complex as a whole, hence if $\bar{S}$ is the total spin, the spin–orbit coupling term to consider in the excitation is $-\zeta_e\bar{L}.\bar{S}$.

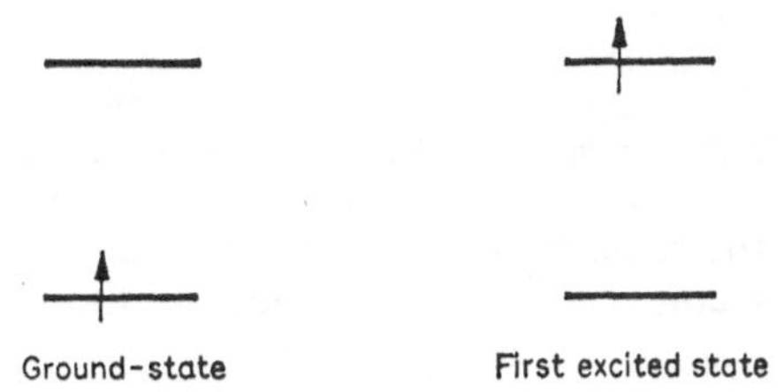

Figure 5-3.

Estimation of Chemical Shifts

Apart from when the valence shell of the atom contains only *s* electrons, chemical shifts are dominated by the paramagnetic contribution, because the diamagnetic contribution of the inner shells will be more or less constant for all the compounds containing the nucleus under discussion. We shall consider an illustrative example which can serve as a prototype for calculations of chemical shifts in general, or for that matter, for ligand field theory calculation of *g*-values.

Chemical Shifts in F_2 and in HF

The first stage of the calculation is to write down the molecular orbitals for the two molecules and these are shown in Figure 5-4, where for simplicity we assume that the fluorine 2*s* orbital are not used significantly in the bonding because of their low energy and high penetration.

The first thing one notices about these diagrams is that the energy gap between the highest filled and lowest unoccupied orbitals is much smaller in F_2 than in HF. Inasmuch as the paramagnetic contribution to the chemical shift varies inversely with excitation energy, we would expect it to be much larger in the case of F_2 than in *HF*.

Another factor to consider is the electronic distributions in the ground and excited states, since the orbital which lead to appreciable

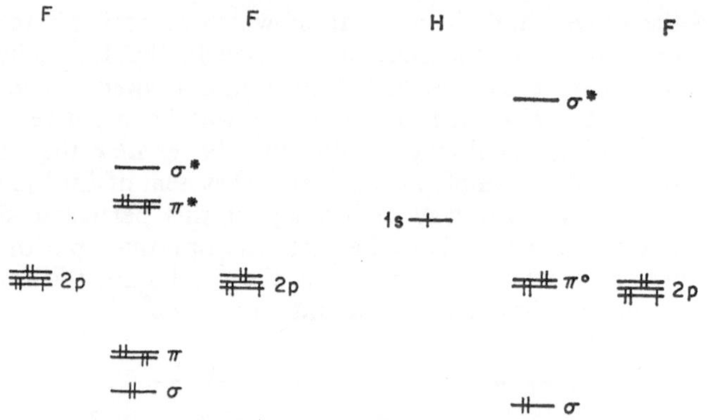

Figure 5-4. Molecular orbitals and electronic configurations in F_2 and in HF

contributions to the chemical shift, must have reasonably high densities near the nucleus concerned. Let us look at the case of *HF* more closely.

The molecular orbitals are:

$$p_x,\ p_y,\ \sigma \text{ and } \sigma^*$$

where

$$\sigma = (\lambda p_z + h)/(1+\lambda^2)^{1/2}$$

$$\sigma^* = (p_z - \lambda h)/(1+\lambda^2)^{1/2}$$

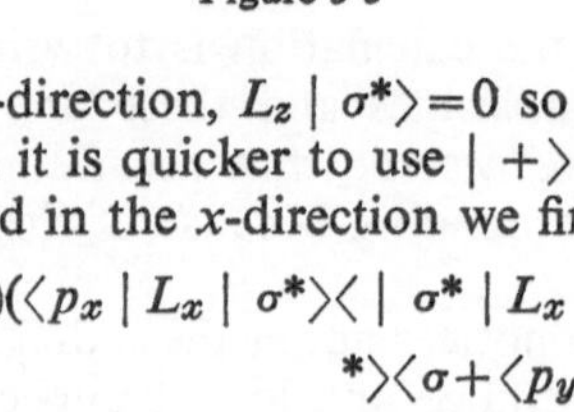

Figure 5-5

If the field is in the *z*-direction, $L_z \mid \sigma^*\rangle = 0$ so that $\sigma_{zz} = 0$. Since p_x and p_y are degenerate, it is quicker to use $\mid +\rangle$ and $\mid -\rangle$ as our basis orbitals, so for the field in the *x*-direction we find,

$$-\sigma_{xx} = 2(2\zeta'_N/\Delta E)(\langle p_x \mid L_x \mid \sigma^*\rangle\langle \mid \sigma^* \mid L_x \mid p_x\rangle$$
$$*\rangle\langle\sigma + \langle p_y \mid L_x \mid \sigma^* \mid L_x \mid p_y\rangle)$$

or

$$-\sigma_{xx} = 2(2\zeta'_N/\Delta E)(\langle + \mid L_x \mid \sigma^*\rangle\langle\sigma^* \mid L_x \mid +\rangle$$
$$+\langle - \mid L_x \mid \sigma^*\rangle\langle\sigma^* \mid L_x \mid -\rangle)$$

where

$$\zeta'_N = \zeta_N/g_N\beta_N$$

i.e.

$$\sigma_{xx} = -(4\zeta'_N/\Delta E)\langle p_z \mid \sigma^*\rangle^2$$

substituting for σ^* the shift due to the four electrons of the "lone-pair p orbitals" is (two in p_x and two in p_y, or two in $\mid +\rangle$ and two in $\mid -\rangle$).

$$\sigma_{xx} = \sigma_{yy} = -(4\zeta'_N/\Delta E)/(1+\lambda^2) \qquad (5.11)$$

The corresponding average value of the chemical shift of the ^{19}F resonance will be,

$$\sigma_p = -(8\zeta'_N/3\Delta E)/(1+\lambda^2) \qquad (5.12)$$

We can proceed along similar lines in the case of F_2, remembering that here all of the electrons are in delocalized orbitals.

Once again, because of the degeneracy, we can use true eigenfunctions of L_z rather than π_x, $\pi_x{}^*$ etc.

The bonding orbitals are then

$$\mid b+\rangle = (\mid +\rangle + \mid +\rangle')\sqrt{2};\ \mid b-\rangle = (\mid -\rangle + \mid -\rangle')/\sqrt{2}$$

and the antibonding orbitals:

$$\mid a+\rangle = (\mid +\rangle - \mid +\rangle')/\sqrt{2};\ \mid a-\rangle = (\mid -\rangle - \mid -\rangle')/\sqrt{2}.$$

Once again $\sigma_{zz}=0$, and the contribution of the two electrons in $\mid a+\rangle$, for example, is given by:

$$\sigma_{xx}(a+) = 2(2\zeta'_N/\Delta E)\langle a+ \mid L_x \mid \sigma^*\rangle\langle\sigma^* \mid L_x \mid a+\rangle$$

The complete expression for σ_{xx}, due to the eight π electrons is:

$$\sigma_{xx} = \zeta'_N(1/\Delta E + 1/\Delta E') \qquad (5.13)$$

where ΔE corresponds to the excitation $\pi^* - \sigma^*$, and $\Delta E'$ to $\pi - \sigma^*$.

A comparison of Eqs. (5.11) with (5.13) leads to two main points. The first is that since we expect that the difference in energy between the π^* and σ^* orbitals will be much smaller than the other differences involved, we would expect the corresponding term to be dominant in determining the paramagnetic shift in F_2 compared with that in *HF*. The second point is that the magnitude of λ determines the magnitude of σ, in *HF*, and since *F* is the most electronegative element, $\lambda > 1$, so that both points lead towards the same conclusion, i.e. that the paramagnetic term should be larger in the case of F_2 than in the case of *HF*. When λ is very large $\sigma_p(HF) \to 0$, which agrees with the limiting case of F^-, where the symmetry and absence of low-lying excited states leads us to expect $\sigma_p = 0$.

5.7. FURTHER EXAMPLES OF CHEMICAL SHIFTS

These simple examples (in Section 5.6) illustrate how the paramagnetic contribution to the chemical shift is largely determined by (a) the energy difference between filled and unfilled molecular orbitals, and (b) the size of the coefficients of the atomic orbitals associated with the nucleus in question, in certain "important" molecular orbitals.

Without going into detail it is not difficult to show how the interplay of these two factors influence the position of a nuclear resonance. For example, when hydrogen is replaced by fluorine in a molecule, six "extra" or "surplus" electrons are introduced as well as the strongly electronegative centre. The overall effect is generally that more antibonding orbitals are introduced and most of them are filled, making or tending to make the energy difference between filled and unfilled orbitals smaller; also in the lowest unfilled (antibonding) orbital, the atomic orbital(s) of the atom of interest will have a large weighting compared with that in the corresponding hydride. This last point is not an absolute rule, of course, but is usually true for simple molecules, when the coefficients of strongly electronegative AO's tend to be large in bonding and small in antibonding orbitals.

Examples:

$$HF \rightarrow F_2 \text{ shift} = -625 \text{ p.p.m. } (^{19}F)$$
$$CH_3H \rightarrow CH_3F \text{ shift} = -77 \text{ p.p.m. } (^{13}C)$$
$$PH_3 \rightarrow PF_3 \text{ shift} = -338 \text{ p.p.m. } (^{31}P)$$
$$(BH_4)^- \rightarrow (BF_4)^- \text{ shift} = -36 \text{ p.p.m. } (^{11}B)$$

These ideas can be extended because, apart from the so-called electron-deficient compounds, the simple hydrides have a set of bonding orbitals, which are filled, and an anti-bonding set which are empty, implying a large energy gap and hence a small contribution to σ_p. Replacement of the hydrogens not only by fluorine atoms, but by any strongly electronegative atoms, such as nitrogen or oxygen, leads to the introduction of π-type interactions and thence to a narrowing of the gap between filled and empty orbitals, implying a corresponding increase in σ_p, relative to the nucleus in the hydride, e.g.

^{14}N chemical shifts: NH_3(0); N_2(−330); NO_2^-(−600); NO_3^-(−350) (p.p.m.)

^{17}O chemical shifts: H_2O(0); NO_2^-(−690); NO_3^-(−420) (p.p.m.)

The effects of electronegativity differences, as extensions of Eq. (5.11) can be virtually isolated in some favourable cases. As we pointed out above, the major factor to consider may be the coefficient of the atomic orbitals in the lowest excited state, i.e. usually this means in the lowest unoccupied orbital, which is generally antibonding. Of course, the coefficients of both orbitals involved in the excitation will be important, but there still seems to be a rough rule that the paramagnetic shift of the nucleus of a strongly electronegative atom goes down when it is combined with a less electronegative element. The opposite holds approximately for the less electronegative atom in this case.

A good example, where the excitation energy is not the major factor producing relative shifts is the isoelectronic series: C_2^{2-}, CN^- and N_2. Assuming that the ^{13}C shift in C_2^{2-} is similar to that in acetylene, we have:

$$^{13}C \text{ shifts } C_2H_2(0); \quad CN^- \, (-105)$$

$$^{14}N \text{ shifts } N_2(0); \quad CN^- \, (+104)$$

The opposite directions of the shifts in going from the homo- to the heteronuclear molecules here can be reasonably ascribed to the different electronegativities of the two elements, in the sense that in CN^-, the higher the energy of an orbital, the greater the weighting of carbon AO's will tend to be, i.e. relative to the more electronegative nitrogen orbitals.

A similar explanation can be put forward for the following series of ^{19}F shifts, which are not necessarily in the order of lowest excitation energies: F_2(0); F_2O (+149); F_3N (+285); F_4C (+491); F_3B (+555); F_2Be (+599); HF (+625).

Finally there is a generalization we can make without having to consider the electronic structure of the molecules in detail, about ^{11}B shifts. When the boron atom is attached to only three ligands or is in an electron-deficient compound such as diborane and is involved in a hydrogen bridge, there is what we can loosely call a non-bonding orbital, which corresponds to an empty boron AO. The presence of this MO means that excitations are smaller than otherwise, and the heavy weighting of boron in it means that we can expect a large ^{11}B paramagnetic shift.

On the other hand, when boron is 4-coordinate, there are no such orbitals, excitations being between bonding and anti-bonding orbitals, only one set of which will contain large coefficients of boron orbitals. The result, in this case is that the paramagnetic shift should be relatively small. A comparison of 3- and 4-coordinated boron supports these deductions:

$B_2H_6 \rightarrow (BH_4)^-$ . . . +55 p.p.m. $BF_3 \rightarrow (BF_4)^-$ . . . +11 p.p.m.

$B(OH)_3 \rightarrow B(OH)_4^-$ . . . +17 p.p.m. $BI_3 \rightarrow BI_3PCl_3$ . . . +63 p.p.m.

With relatively few starting principles, such as the effects of electronegativity differences on the shapes of MO's, and the greater π-bonding capacity of the first row elements, most chemical shifts can be rationalized in terms only of the paramagnetic contribution, implying that as expected, differences in the diamagnetic part are usually small, for a given nuclear spin.

This is not the case, however, with protons, due to the absence of orbital angular momentum in the ground-state of the hydrogen atom, at least in the simple terms in which we are speaking, so it is expedient to discuss proton chemical shifts separately.

5.8. PROTON CHEMICAL SHIFTS

The absence of orbital angular momentum in the ground-state of the hydrogen implies that the paramagnetic term for proton chemical shifts, can arise only from circulations induced on other atoms. The effects on a nucleus can be classified into those which are local and others which are more remote. By local effects we mean those due to circulations induced in the electron density in the atomic orbitals of that atom, and by non-local effect we mean those due to induced currents on other atoms or in bonds, which may be delocalized.

In the first place, then, the local paramagnetic effect is zero, so we might expect the diamagnetic contribution, which is probably rather unimportant in the chemical shifts of other nuclei, to dominate proton chemical shifts. Since the local diamagnetic shift depends on the electron density around the nucleus, we would expect a straightforward correlation between chemical shift and the electronegativity of the adjoining atom. The figures in Table 5.8(1) show that this is not so, for the shift in HF is approximately the same as that in H_2, for example, and the proton resonance in methane (and in H_2S) is at lower field than that in HCl.

Table 5.8(1) Proton Chemical Shifts in Some Simple Molecules (in ppm)

"Bare" proton	H_2	HF	HCl	HBr	HI	H_2O	H_2S	NH_3	CH_4
−27.6	0	0.3	3.2	7.2	16.1	2.2	2.7	2.9	2.8

It is clear that, since the electron density in the hydrogen AO in HF is very much less than that in H_2, there must be some other factor to consider, other than the Larmor precession of the "local" electrons. A clue to the solution of this problem is given by the fact that the range of chemical shifts in these widely different compounds is an order of magnitude smaller than those found in the resonance of other nuclei. This is because the total electron density around a proton is never large, i.e. is always between 0 and 2, so that the diamagnetic effect is necessarily small. An important consequence of this is that though the fields at the proton due to induced circulation elsewhere in the molecule are small, they are, nevertheless, often comparable in magnitude to the local diamagnetic field.

There are some cases where chemical shifts do seem to reflect the charge density on a hydrogen atom without other complications, e.g. LiH (+26.5)*; H_2 (0); H^+ (−26.5); and in organic compounds positive ions tend to give resonances at low field, whereas, in negative ions the proton resonances tend to be at high field, corresponding to less and greater electron density in the hydrogen atomic orbitals respectively, e.g. $C_5H_5^-$ (+1.73); C_6H_6 (0); $C_7H_7^+$ (−1.87); $C_8H_8^{2-}$ (+1.58), where we have taken the protons in benzene as standard.

In organic compounds in general there is a rough correlation between the expected charge density near a proton and its chemical shift, for example, the methylene protons in C_2H_5OH or in C_2H_5Cl resonate at a lower field than the methyl protons. There are, however, a number of well-defined exceptions to this rule, as we can see from the following order of resonant fields for a constant frequency:

low field CH_2O C_6H_6 C_2H_4 CH_3OCH_3 C_2H_2 cyclo-C_6H_{12} C_2H_6 high field

← about 10 p.p.m. →

In order to explain why, for example, the shifts are not in the same order as the acidities of the compounds it is necessary to look at what we have called non-local effects. The main thing to understand about the fields at a proton due to currents induced in other parts of the molecule, is that for any effect to be observable, in solution, the induced currents must be anisotropic, i.e. for different orientations of the molecule in the applied field, the magnitudes of the induced currents in these other atoms/bonds must be different. We can see why this condition has to be satisfied from the following example where the induced current on X has a different value when the bond H–X is parallel, than when it is perpendicular, to the applied field.

When the applied field is parallel to H–X, the magnetic effects of the induced current on X can be represented approximately by a point dipole $\mu_{\parallel}B$. For the orthogonal positions the corresponding moment is $\mu_{\perp}$ (see Figure 5-6).

The average field at the proton due to its neighbour will therefore be $-(1/3r^3)(2\mu_{\parallel}-2\mu_{\perp})B$, which vanishes when $\mu_{\parallel}=\mu_{\perp}$. Here the induced moments on the atom X may be due to a combination of diamagnetic and paramagnetic effects, however, one generally expects that the diamagnetic part will be more or less isotropic, since it involves the Larmor precession of the electron density associated with X about

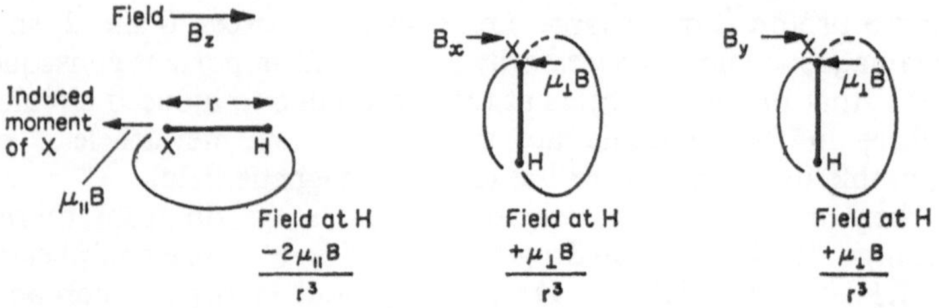

Figure 5-6.

its nucleus, and this is not sensitive to the direction of the field, because it will be approximately spherically symmetrical. Differences in paramagnetic currents are liable to be very large, on the other hand, and are often strongly dependent on orientation, so it is these which largely determine not only the shifts of the nucleus of X, but also the neighbouring bond magnetic anisotropy effect on the proton attached to X. Protons further removed from X are unlikely to be affected appreciably by these currents because the induced field falls off rapidly, i.e. with r^{-3}, with distance.

The proton resonances of the halogen acids and of acetylene occur at rather higher fields than expected at first, and this can be ascribed to paramagnetic circulation adjacent to the protons.

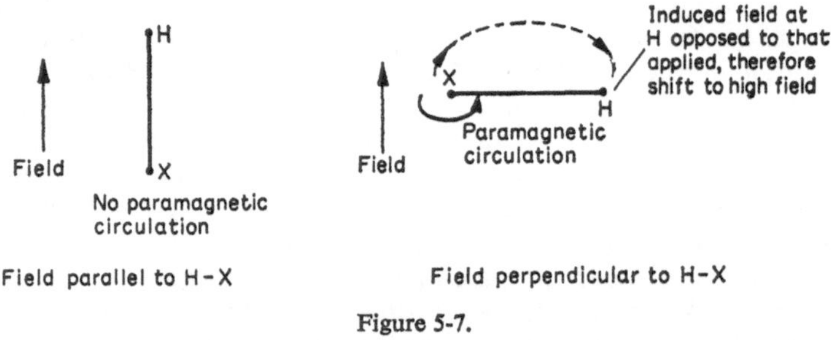

Figure 5-7.

The diamagnetic part of the circulation is virtually the same in both cases. The effect is at an optimum with linear molecules since then there is a maximum difference between $\mu_{\parallel}$ and $\mu_{\perp}$, because excitations involving other bonds will give low field shifts when field is parallel to H–X, i.e. reducing magnitude of the average.

Ring Currents

The proton resonance of aromatic compounds is at much lower field than expected for sp^2 carbon systems, and this has been neatly explained in terms of the Larmor precession of the π-electrons around the periphery of the rings. Thus when there is an annular system, there is the possibility of an induced current flowing around the circuit, and in systems in which there are no low-lying excited states, this will be in the sense so as to oppose the applied field. The magnitude of the current is given by the Larmor theorem, i.e. the electrons will be precessing at a rate $eB/2mc$, so knowing the radius and area of the ring we can calculate the current flowing and its magnetic effects.

For distances further from the centre of the ring than its radius, it is a satisfactory approximation to replace the current by a suitable point dipole, situated at the centre of the ring, for calculation of magnetic effects. The corresponding fields at the protons outside the ring will depend on orientation, and in solution there will be a resultant "de-shielding" effect, in contrast to the local Larmor precession of the electrons in the hydrogen AO which produces a "shielding" effect.

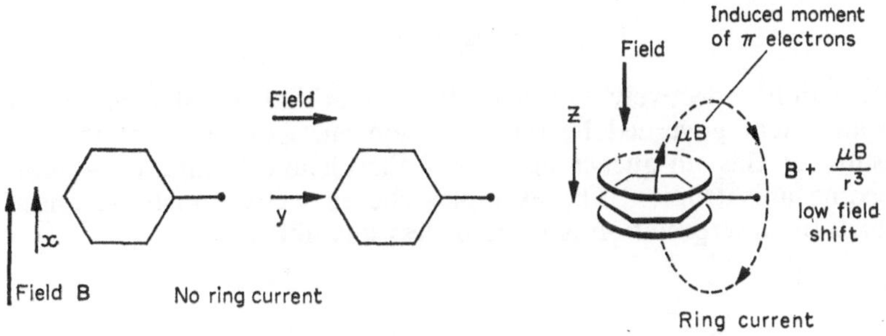

Figure 5-8. Ring current in benzene.

Support for the ring current idea can be drawn from

(a) Quantitative calculations for a large variety of polynuclear alternant hydrocarbons, such as naphthalene, phenanthrene etc., when each hexagon of carbon atoms is identified magnetically with a benzene ring, so that in the point dipole approximation naphthalene, for example, is represented by two magnets, one at the centre of each 6-membered ring. Each polynuclear hydrocarbon is treated like a network in which the resultant current flows around the periphery.

(b) Qualitatively, in some alternant systems, there are protons inside, as well as outside the ring, and as expected these show large high-field shifts (see Figure 5-9).

In this chapter we have tried to illustrate how the theories of g-tensors and of chemical shifts are linked, i.e. the formulae are analogous and both depend on the extent of interaction of the appropriate spin and the electronic orbital angular momentum. Such

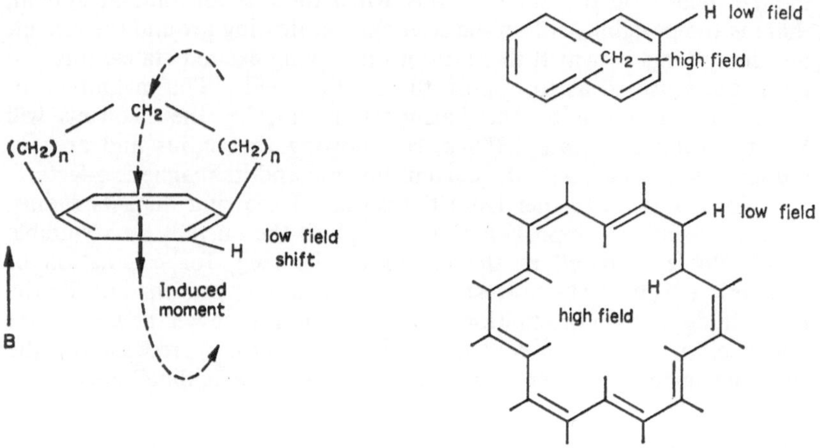

Figure 5-9.

interactions effectively "un-quench" the orbital angular momentum to an extent governed by the excitation energies. One of the main results of this un-quenching is that the chemical shift, or g-tensors become anisotropic. The examples chosen were, hopefully, ones in which the interpretation is more or less unambiguous.

Chapter 6

The Theory of Coupling Constants

6.1. INTRODUCTION

As we have implied already, when we observe the spin of a particle it is invariably in an electromagnetic environment. In particular it is surrounded by other particles, at various distances, which may or may not have spins, and it is the interactions between the particle's spin and other spins which we shall examine in this chapter. To a good approximation these interactions, which are responsible for hyperfine splitting of the resonance spectrum, are independent of the applied field, so we do not have to introduce the field into the discussion. Also since the electronic orbital angular momentum is effectively quenched in the molecules and radicals which we shall consider, we shall not be including the interactions between spins and the orbital motion of the electrons. Furthermore most of our examples will be about radicals or molecules in solution, when the anisotropic dipole–dipole interaction between the spins averages to zero, so in these cases we shall consider only the Fermi contact term in trying to rationalize the empirical data.

Before we look at the matter in more detail, it is worth noting that hyperfine splitting occurs in ESR spectra due to the delocalization of the electrons, in the first place, and to the repulsions of the electrons, which tends to keep electrons of opposite spins as far apart as possible, in the second. The situation is the same in NMR spectra, except that the calculations are one degree further in complication. In both cases coupling constants are determined by electron–nuclear spin interactions.

6.2. HYPERFINE SPLITTING IN ESR SPECTRA

Hyperfine Coupling in Atoms

Since atoms are always spherically symmetrical, the dipole–dipole interaction between an odd electron and the nuclear moment will be zero to the first order, so the coupling energy will be determined by the Fermi contact interaction. It follows from this that any observed splitting in the resonance spectrum of an atom is due to the odd electron density at the nucleus. In the orbital approximation this implies that the atomic orbital of the odd electron must have a finite value at the nucleus, and must therefore have the character of an "*s*" type of orbital. The simplest example is, as one might expect, that of the hydrogen atom, for then there is only a 1*s* orbital involved and we know the ground-state wave function exactly.

The general form for a 1*s* orbital is:

$$\psi = (Z^3/a_0^3)^{1/2} \exp(-Zr/a_0)$$

where Z = atomic/effective atomic number and a_0 = Bohr radius of the H atom. For a hydrogen-like atom, substitution of values for the various constants leads to the following formula for the first-order interaction between the odd electron and the nucleus:

$$\begin{aligned} a_{\mathrm{H}} &= (508.5)\, Z^3 && \text{(for hydrogen)} \\ a_X &= (508.5)\, Z^3\, (\mu_X/\mu_{\mathrm{H}}) && \text{(for an atom } X\text{)} \end{aligned} \tag{6.1}$$

where μ_X = magnetic moment of nucleus X.

In the case of hydrogen atoms ($Z=1$) produced by the irradiation of water or paraffins, the observed splitting is in exact agreement with that predicted from the above "purely" theoretical formula Eq. (6.1).

Considerable hyperfine splittings are observed, however, in the ESR spectra of, for example, the halogen atoms, in spite of the fact that, in simple terms, the odd electron occupies a *p* orbital, which, being of the form $\bar{r}f(r)$, vanishes at the nucleus. These are situations in which the orbital approximation is inadequate, and in order to explain the coupling constants observed, we have to introduce configurational interaction. In the case of fluorine, for example, the electron repulsions lead to a mixing of the configurations $1s^2 2s^2 2p^5$ and $1s^2 2s^1 2p^5 3s^1$ (amongst others), and this is what leads to a prediction of observable hyperfine coupling. In slightly different terms, when we start from the orbital approximation we say that the odd electron polarizes the spins of the other, initially paired, electrons in such a way that spin density appears in places where naively one might not expect any. Generally the induced spin density is in the same sense as the spin of the odd electron in its immediate vicinity, and of the opposite sense in more remote parts of the atom or molecule. Whether the spin density induced at the nucleus is positive or negative, i.e. is in the same or in the opposite sense to that of the odd electron, depends mainly on the repulsion integrals, which in turn depend on the extent to which the odd electron moves in the space defined by the filled/empty *s* orbitals.

Thus in the case of fluorine, above, the spin density induced in the 2*s* orbital will depend on the balance of two repulsions one between a 2*s* and a 2*p* electron, and the other between a 3*s* and a 2*p* electron. It will be in the opposite sense to that induced in the 3*s* orbital.

It must be remembered here that this type of discussion only has meaning if we are talking in the context of the orbital approximation, i.e. we start our calculations by assigning the electrons to appropriate atomic orbitals and then look to see how to improve on this approximation, since it accounts for electron repulsions only very crudely and does not admit any electron correlation, i.e. "the electrons tend to keep out of each other's way". This applies equally to molecular orbital theories, which are only useful inasmuch as they represent, without much refinement, a reasonably (or unreasonably!) good picture of the situation under discussion.

6.3. MOLECULAR ORBITAL THEORY OF COUPLING CONSTANTS IN FREE RADICALS

The orbital approximation is particularly attractive to anyone trying to predict or explain coupling constants, because all one has to do, in the first place at least, is to find the form of the molecular orbital which contains the odd electron and hence the value of its amplitude at the nucleus in question.

The commonest, simplest and most useful approach is the Hückel or extended-Hückel method, which is the most simple form of the LCAO MO theory. The first aim is to find the coefficients of the various atomic orbitals in the molecular orbital assigned to the odd electron. The coupling constant is then given by the product of the expression of Eq. (6.1) and the square of the coefficient(s) of the atomic orbital(s) associated with the nucleus. In the case of hydrogen the expressions is:

$$a_{\mathrm{H}} = 508.5 Z^3 c_{\mathrm{H}}^2 \qquad (6.2)$$

where c_{H} is the coefficient of the hydrogen AO and Z the effective atomic number of hydrogen in a molecule, which is often taken to be 1.2 since the orbital is somewhat contracted there, according to *ab initio* calculations for H_2^+ and H_2.

β-Proton Coupling Constants

In most simple aliphatic radicals the odd electron is more or less localized in a particular carbon atomic orbital. The splitting due to protons on the atom adjacent to this position of maximum spin density can be sensibly discussed in simple MO terms. This is also true for the methyl proton coupling constants in methyl-substituted aromatic radical-ions.

The coupling of these β-protons arises from the direct delocalization of the odd electron by what is termed "hyperconjugation". We shall consider two examples, one a π-type of radical, in which the odd electron is mainly in a p orbital, and the other a σ-radical in which the odd electron is largely associated with a hybrid orbital.

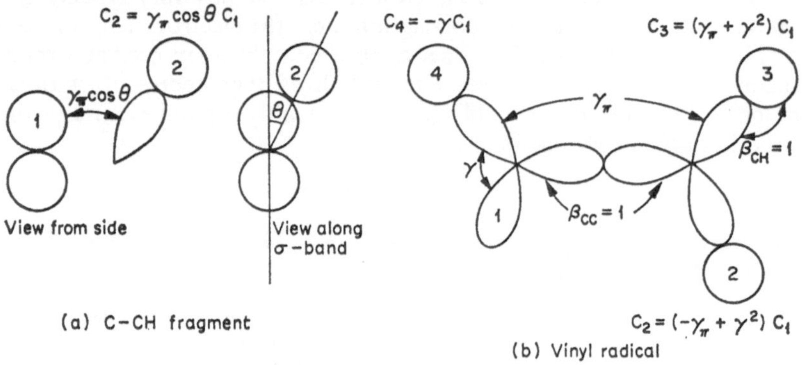

Figure 6-1. Orbitals and their coefficients in two examples of β-coupling.

In these examples, given in Figure 6-1, we express the resonance integrals in terms of β_{CH} and we neglect all those which seem to have little bearing on the results. Equating all of the coulomb integrals, for simplicity, we find that with this scheme of parameters in both cases the odd electron goes into a non-bonding orbital. This makes it very easy to calculate the coefficients, and those which are non-zero are shown in the figure.

Looking first at example (a) we see that the coupling constant should vary with $\cos^2 \theta$, where θ is the dihedral angle, assuming that γ_π is small. This is what is observed and in fact, β-coupling constants in a large variety of alkyl and cycloalkyl radicals can be calculated with reasonable accuracy using $\gamma_\pi = \frac{1}{4}$ and $Z = 1.2$, in Eq. (6.2).

Example: compare the splitting in cyclopentyl ($\theta = \pi/6$), with that in isopropyl (average $\cos^2 \theta = \frac{1}{2}$). The theoretical ratio is

$$a(\text{isopropyl})/a(\text{cyclopentyl}) = \tfrac{1}{2}/\tfrac{3}{4} = \tfrac{2}{3}$$

The observed ratio is

$$24.68/35.12 = 0.68$$

Incidentally, the α-protons in this example are in the nodal plane of the p orbital and therefore in simple MO terms we would expect no splitting from them. The observed α-splittings are, however, quite large, usually being more than 20 G. This can only be explained if we introduce other configurations into the discussion, in order to allow for electron repulsions in a more refined way. It is convenient to postpone this complication till the next section.

In the second example, it is clear, using our parameterization, that $a_{\text{cis}} < a_{\text{trans}}$ since $c_2^2 < c_3^2$. This is in agreement with the observed values $a_{\text{cis}} = 34$ g, $a_{\text{trans}} = 68$ G, confirming the importance of hyperconjugation as a mechanism for delocalizing spin.

6.4. THE VALENCE-BOND THEORY OF COUPLING CONSTANTS–SPIN POLARIZATION–NEGATIVE SPIN DENSITY

In the above examples we considered only the simple delocalization of the odd electron in trying to explain coupling constants. This mechanism is not very important in determining the hyperfine splitting by protons which are in the nodal plane of the odd-electron orbital, even allowing for molecular vibrations out of the plane of symmetry. In order to explain the hyperfine interaction of a proton in, say, the methyl radical or in the benzene negative ion, it is convenient to introduce the idea of spin polarization. This is best illustrated in terms of valence bond theory.

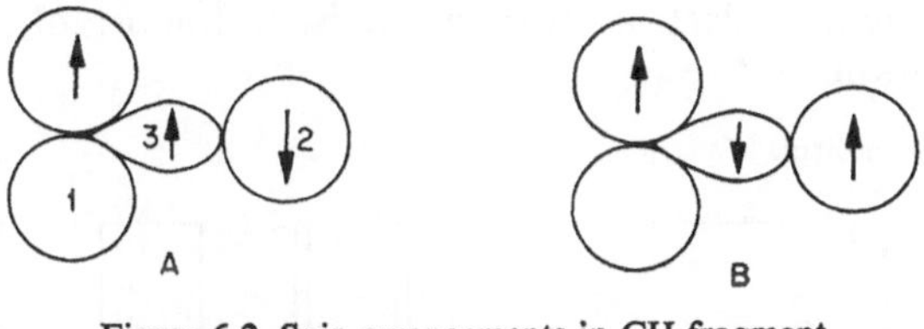

Figure 6-2. Spin arrangements in CH fragment.

Let us consider a three orbital system as in Figure 6-1a, i.e. one carbon AO in which the odd electron resides, and two other atomic orbitals which make a C–H bond. We may regard the system either as a C–CH fragment as in the previous section, or as a CH fragment (see Figure 6-2).

If π- and σ-systems were truly isolated from each other the spin states *A* and *B* would occur with equal weighting in the VB wavefunction of the fragment and the net spin at the proton would be zero. However, simple application of Hund's rule tells us that electrons in different orbitals of the same atom tend to be of the same spin. This is because if their spins are parallel, the spatial part of their combined wavefunction must be anti-symmetric (Pauli principle) so that they tend to keep out of each other's way more than if they were paired, implying a smaller repulsion energy. The result of this is that spin state *A* is preferred over *B*, so that there is a net spin associated with the hydrogen AO which is opposite in sign to that of the radical fragment as a whole. This is what is meant by the term "negative spin density".

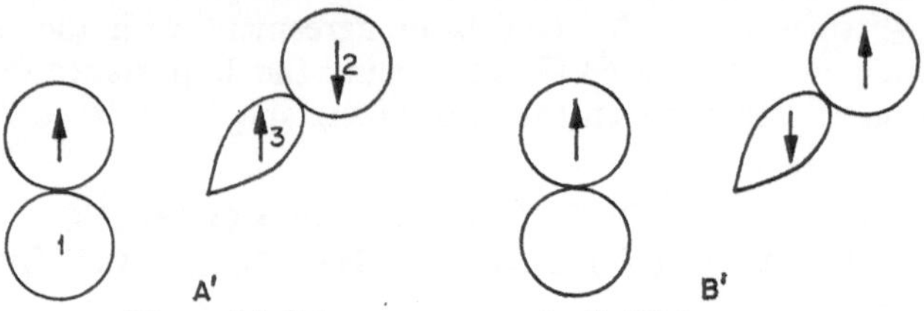

Figure 6-3. Spin arrangements in C–CH fragment.

When we consider the analogous spin states for β-coupling constants the result is different (see Figure 6-3), since electrons in orbitals on different atoms tend to pair up to form a bond, in this case when $0<\theta<\pi/2$. So we expect, from this simple model, that splittings should be positive, i.e. the spin density at the β-proton has the same sign as the fragment as a whole. In the case of β splittings, it is the forming of "partial" bonds which plays the major role, rather than electron repulsions (i.e. correlation). We associated this mechanism with the term, electron delocalization, whereas in connection with α splittings, where electron correlation is all-important, we talk of spin polarization.

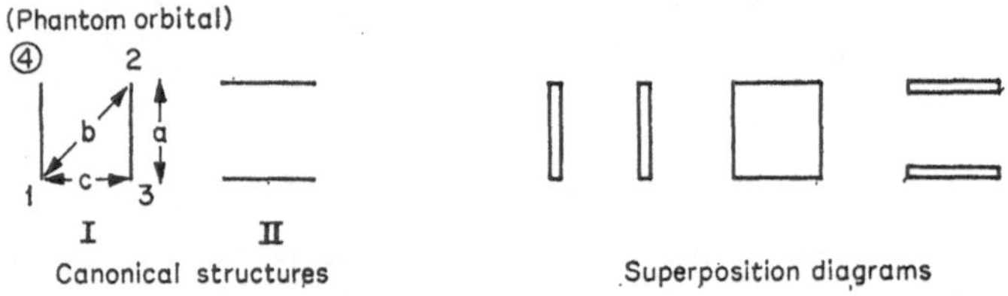

Figure 6-4. Rumer diagrams for a four orbital system.

We have, of course, oversimplified the situations, since we have only considered one of the non-bonded interactions in each case. To be a little more explicit let us look at these problems in simple VB terms.

We make it into a four orbital problem by introducing a "phantom" orbital which enables us to complete a set of canonical structures (in this case two) each containing two bonds (see Figure 6-4).

The VB wavefunction is given by the linear combination:

$$\Psi_{VB}=c_{I}\psi_{I}+c_{II}\psi_{II}$$

We solve for the coefficients using the usual variational method and the matrix elements, such as $\int \psi_I(\mathscr{H}-E)\,\psi_I\,d\tau$, are found from the corresponding superposition diagrams and the formula

$$1/2^{n-r}(x + J\text{'s on same islands separated by odd number of bonds}$$
$$-\tfrac{1}{2}J\text{'s on different islands}$$
$$-2J\text{'s on same islands separated by even number of bonds})$$

where x = coulomb integrals $- E$

n = number of bonds in the canonical structure

r = number of islands in superposition diagram, an island being a closed figure.

The secular equations are:

$$(x+a-\tfrac{1}{2}c-\tfrac{1}{2}b)\,c_{\mathrm{I}}+\tfrac{1}{2}(x+a-2b+c)\,c_{\mathrm{II}}=0$$
$$\tfrac{1}{2}(x+a-2b+c)\,c_{\mathrm{I}}+(x+c-\tfrac{1}{2}a-\tfrac{1}{2}b)\,c_{\mathrm{II}}=0$$

Now since $c_{\mathrm{I}} \gg c_{\mathrm{II}}$, $x+a-\tfrac{1}{2}c-\tfrac{1}{2}b \approx 0$ so that:

$$c_{\mathrm{II}}/c_{\mathrm{I}}=\tfrac{1}{2}(c+b)/(a-c)$$

The spin states of the two canonical structures are:

for I $\alpha\alpha\beta-\alpha\beta\alpha$
(in the order 1, 3, 2)
for II $\beta\alpha\alpha-\alpha\beta\alpha$

In terms of spin the wavefunction of the fragment is α, therefore:

$$c_{\mathrm{I}}\alpha\alpha\beta-(c_{\mathrm{I}}+c_{\mathrm{II}})\,\alpha\beta\alpha+c_{\mathrm{II}}\beta\alpha\alpha$$

The α-spin density on the proton is:

$$-c_{\mathrm{I}}^2+(c_{\mathrm{I}}+c_{\mathrm{II}})^2+c_{\mathrm{II}}^2$$

i.e.

$$=2c_{\mathrm{II}}(c_{\mathrm{I}}+c_{\mathrm{II}}) \approx +2c_{\mathrm{I}}c_{\mathrm{II}}$$

The integral "a" will be large and negative, since it corresponds to a σ-bond, b will be small but c will be positive, in the case of an α proton, because it corresponds to a repulsion, and negative in the β proton case. We would therefore expect α proton splittings to be negative, because c_{I} and c_{II} have opposite signs. In contrast β splittings should be positive.

As we have already said, the simple MO wavefunction for the fragment, which we can represent by $(a\sigma\sigma)(\alpha\beta\alpha-\alpha\alpha\beta)$ leads to no spin polarization and to positive spin densities only where the coefficient of the orbital in "a" has non-zero values. A better ground-state wavefunction is obtained if we take linear combinations of this function with excited MO states and utilize the variational principle. The excited state which leads to spin polarization in the σ-bond is $(a\sigma\sigma^*)(2\beta\alpha\alpha-\alpha\beta\alpha-\alpha\alpha\beta)$ and the ultimate result is that an α-proton coupling constant of about -20G is expected, in agreement with the VB predicted value and that obtained experimentally.

The CH bond interacts most strongly, of course, with the nearest part of the π-electron system, so we expect that the splittings from either α- or β-protons should be roughly proportional to the odd electron density on the nearest carbon p_π orbital, i.e.

$$a_{\rm H}=Q^{\rm H}{}_{\rm CH}\rho_{\rm C} \tag{6.3}$$

where $Q^{\rm H}{}_{\rm CH}$ is a constant and $\rho_{\rm C}$ the appropriate spin density. This formula holds reasonably well provided that distortions of bond angles are not too large.

Table 6.I Coupling constant of protons attached to sp^2 carbon in aromatic radicals

Radical	CH	$C_6H_6^-$	C_5H_5	C_7H_7	$C_8H_8^-$
Q (obs.)	-23.04	-22.5	-29.9	-27.4	-25.7

(Signs in solutions have to be inferred.)

Spinstates

I $(\alpha\beta\alpha-\beta\alpha\alpha)$

II $(\alpha\beta\alpha-\alpha\alpha\beta)$

Figure 6-5.

The usual procedure in explaining the ESR spectra of aromatic radicals is to calculate the spin densities in the carbon p AO's then to use Eq. (6.3) to obtain the theoretical splittings. Sometimes the distance between the outermost lines of the spectrum is greater than Q, and when this is so the implication is that there must be positions in the electron system on which there are negative spin densities. The most simple example of this is that of the allyl radical $CH_2CH:CH_2$.

According to simple MO theory the spin densities in this radical should be $\frac{1}{2}$; 0; $\frac{1}{2}$ so that the spectrum should be either a quintet (1 : 4 : 6 : 4 : 1) due to four equivalent protons, or nine lines (1 : 2 : 1 : 2 : 4 : 2 : 1 : 2 : 1) overlapping to a certain extent due to two pairs of equivalent protons, allowing for the bent nature of the radical. This is not what is observed. The VB approach provides a better description in this case for the two obvious canonical structures I and II, (see Figure 6-5) must appear with equal weights in the VB wave-function for the ground-state (i.e. $\psi_{\rm VB}=c_{\rm I}\psi_{\rm I}+c_{\rm II}\psi_{\rm II}$), so that the spin wavefunction is represented by: $(1/\sqrt{6})2\alpha\beta\alpha - \alpha\alpha\beta - \beta\alpha\alpha)$. The α-spin densities are therefore:

$$\tfrac{1}{6}(4+1-1)=\tfrac{2}{3} \text{ on the end positions}$$

$$\tfrac{1}{6}(-4+1+1)=-\tfrac{1}{3} \text{ on the centre position}$$

The spectrum should therefore have twice as many lines as that predicted from simple MO theory. It is interesting to compare the predictions with experiment for we can see that the observed coupling constants are about half-way between those predicted from these simple

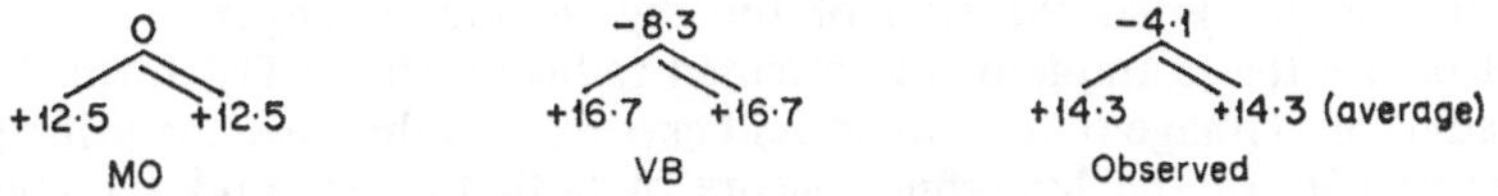

Figure 6-6.

theories taking $Q=25\text{G}$. As a general rule negative spin densities occur whenever simple Hückel theory predicts zero spin density on a position close to one of high spin density. One class of compounds in which this situation occurs is that of the odd-alternant hydrocarbons, another example being the benzyl radical.

6.5. ANISOTROPIC COUPLING CONSTANTS IN SOLIDS

The components of the hyperfine coupling tensor have been measured for a few radicals trapped in solids. If the odd electron is in an *s*-type of atomic orbital, all of the diagonal elements of the tensor will be the same when the orbital is associated with the same atom as the nucleus, but for other nuclei the tensor will be anisotropic, due to the dipole–dipole interaction. The coupling will consist of an isotropic part, "*a*", from the Fermi contact term, and an anisotropic part determined by the function

$$\bar{I}.\bar{S}/r^3 - 3(\bar{I}.\bar{r})(\bar{S}.\bar{r})/r^5$$

where $\bar{r}$ is the vector joining the nucleus to the electron.

The diagonal elements of this anisotropic part are given by formulae such as:

$$T_{zz}=\text{constant }(1-3\cos^2\theta)/r^3$$

where θ is the angle between the z-axis and the line joining the nucleus to the electron (see Figure 6-7).

Similar formulae apply for the other two diagonal components. One of the useful pieces of information which can be deduced from

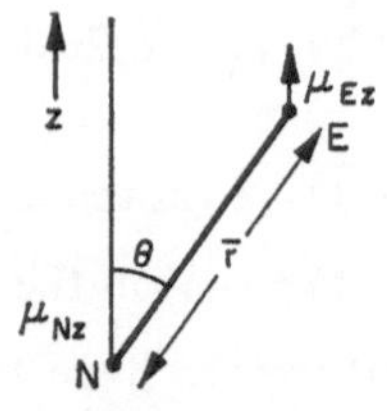

Figure 6-7.

single crystal studies is the sign of the coupling constant. This is possible because the anisotropic component is positive in sign only when $\cos^2\theta > \frac{1}{3}$, so the sign of the dipole term changes at $\theta = 55°$. Let us use the example of the oriented radical $\dot{C}H(CO_2H)_2$ formed by irradiation damage in a malonic acid crystal. In this case the principle components of the hyperfine tensors of both the ^{13}C and ^1H nuclei are known.

In the first place, the applied field determines the direction along which measurements are made, and by making observations at a sufficient number of intermediate points, the relative signs of the principle components can be found, and from them the magnitude of the isotropic part of the coupling constants.

The odd electron is presumably in a carbon $2p$ orbital which we will take as defining the z-direction for the radical. We shall look at the ^{13}C splitting first because it is easier to see; when the field is along the z-axis almost all of the odd-electron density is within the cone $\theta = 55°$ (see Figure 6-8) and therefore the anisotropic coupling is positive. On the other hand when the field is along an axis perpendicular to this, most of the electron spin density is outside the limit

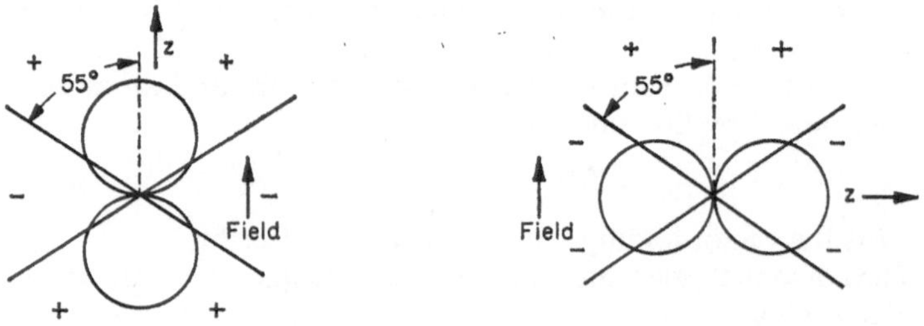

Figure 6-8. ^{13}C anisotropic coupling.

$\theta = 55°$; so the dipole–dipole coupling here should be negative. The observed coupling tensor is nearly axially symmetric and from the above reasoning we would expect the two components which are nearly equal to have negative contributions from their anisotropic coupling with the odd electron (call these T_1' and T_2'). Since we already know the relative signs and the magnitudes, we can write down the principle components with their signs, in the usual way, i.e. writing the scalar coupling separately

$$a_C = +92.6; \quad T_1' = -50.4; \quad T_2' = -59.8; \quad T'_3 = +120.1$$

For completeness we write down the components as they are found from experiment, i.e. before all the above reasoning:

$$T_1 = +42.2, \quad T_2 = +32.8, \quad T_3 = +212.7$$

The principle axes of the coupling tensor of the proton are parallel to those of the ^{13}C coupling tensor and we can treat the observed values in an analogous way. They may be either all positive or all negative and their values are (29, 91, 61), at first sight, but consideration of Figure 6-9 leads us to the set

$$a_H = 660.3; \quad T_1' = +31.3; \quad T_2' = -30.7; \quad T_3' = -0.7$$

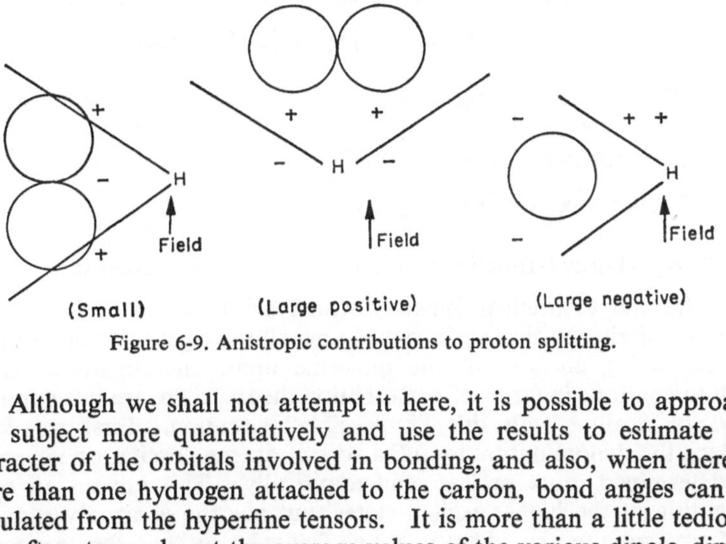

Figure 6-9. Anistropic contributions to proton splitting.

Although we shall not attempt it here, it is possible to approach this subject more quantitatively and use the results to estimate the character of the orbitals involved in bonding, and also, when there is more than one hydrogen attached to the carbon, bond angles can be calculated from the hyperfine tensors. It is more than a little tedious, very often, to work out the average values of the various dipole–dipole interactions!

6.6. NUCLEAR SPIN–SPIN COUPLING IN SOLUTION

The theory of nuclear spin–spin interactions is more complex than that of hyperfine coupling in ESR because it involves a perturbation of the second order rather than one of the first order. One result of this is that the splittings observed in NMR are of the order of 10^6

times smaller than those in ESR so they are measured in c/s (Hz) rather than Mc/s (MHz).

The interactions between two nuclear moments can be classified as follows:

(a) Dipole–dipole interactions between them or between the two nuclei and the electron spins.
(b) Interactions between the nuclear spins and the electron orbital motion.
(c) Fermi contact interaction between the nuclear and electron spins.

For the lighter nuclei in solution only the last of these need be considered since the corresponding perturbation energy gives the largest contribution to the magnetic interaction between the two nuclei.

The interactions which determine nuclear spin–spin coupling constants are the same as those which determine the splittings in ESR spectra, but since we are usually dealing with molecules which have no unpaired electrons the coupling energy is given by the formula:

$$E_{\text{pert}} = -\sum_n \langle 0 \mid \mathscr{H}' \mid n\rangle \langle n \mid \mathscr{H}' \mid 0\rangle / \Delta E_n \tag{6.4}$$

where $\mathscr{H}' = \sum_r \sum_N K_N \bar{S}_r \,.\, \bar{I}_N \delta_r(N)$

K_N = constant for nucleus "N"

$\sum_r$, sum over electrons; $\sum_N$, sum over nuclei

$\delta_r(N)$ = Dirac δ-function; n is a label for the nth excited state.

The first theoretical difficulty which has to be faced is that this series is infinite, so that we have to know all the eigenfunctions and the corresponding energies of the molecule under investigation. If we wish to continue in terms of perturbation theory, there are two common ways of dealing with this "impossible" situation. First we could replace the denominators by ΔE a sort of average excitation energy to be determined more or less semi-empirically. This implies an overweighting of the higher-excited states, but has the great advantage in that we can use the identity $\sum \mid n\rangle \langle n \mid \equiv 1$; so that the expression reduces to a single term depending only on the ground-state, i.e.

$$E_{\text{pert}} = -\langle 0 \mid \mathscr{H}'^2 \mid 0\rangle / \Delta E \tag{6.5}$$

Secondly we could consider the terms involving the lowest excited states only, and neglect all the others, i.e. underestimate the effects of the higher excited states. It is only practicable to use this second approach when it is relatively easy to write down some sort of approximate wavefunction for the excited states of low energy, such as in simple MO theory, where it is as easy to find the excited states as it is to find the ground-state wavefunction.

Physical Picture of the Coupling

We can get a good idea about the mechanism of the indirect coupling of nuclear moments by considering the most simple molecule, H_2. Suppose the spatial parts of the hydrogen atomic orbitals are a and b and that our convention is that instead of labelling the functions of electrons the order in a product tells us which functions are to be associated with which electron, e.g. ab means electron 1 in a, electron 2 in b, and so on. In these terms the VB ground-state wavefunction is

$$|\,0\rangle=(ab+ba)\,(\alpha\beta-\beta\alpha)/\sqrt{2}$$

and the first excited state with a zero z-component of spin (electronic) is

$$|\,1\rangle=(ab-ba)\,(\alpha\beta+\beta\alpha)/\sqrt{2}$$

One way of allowing for the nuclear spins is to find the best linear combination of these wavefunctions, i.e. using the variational principle. Thus a better wavefunction for the actual state of the molecule is

$$\Psi=c_1\,|\,0\rangle+c_2\,|\,1\rangle$$

i.e. $$\Psi=[(c_1+c_2)\,(a\alpha b\beta-b\beta a\alpha)+(c_1-c_2)\,(a\beta b\alpha-b\alpha a\beta)]/\sqrt{2} \qquad (6.6)$$

Thus if c_1 and c_2 have the same sign, α-spin will be preferred in a and β-spin in b. We can estimate c_2, which is naturally small, using first-order perturbation theory, i.e.

$$c_2=-\langle 1\,|\,\mathscr{H}'\,|\,0\rangle/\Delta E$$

Now

$$\mathscr{H}'=+K_N[\delta_1(A)\,\bar{S}_1.\bar{I}_A+\delta_2(A)\,\bar{S}_2.\bar{I}_A+\delta_1(B)\,\bar{S}_1.\bar{I}_B+\delta_2(B)\,\bar{S}_2.\bar{I}_B]$$

The most simple way to continue is to assume that the two nuclei are not identical so that we could observe their coupling in their NMR spectrum. We only have to consider the z-components of the spins and the above triplet wavefunction. The expression for c_2 is then

$$c_2=-k(a_0{}^2I_{Az}-b_0{}^2I_{Bz})^2/\Delta E \qquad (6.7)$$

where a_0=value of a at nucleus A

b_0=value of b at nucleus B

We can see immediately that if the nuclear spins are parallel then $c_2=0$, so no effects would then operate. When the spins are antiparallel there would be local unbalancing of the electron spins, one type of electron spin being preferred at one end of the bond and the other at the other end. From Eqs. (6.6) and (6.7) above it can be deduced that α-electron spin is preferred near to A when this nucleus has β-spin (i.e. when $m_{Az}=-\frac{1}{2}$).

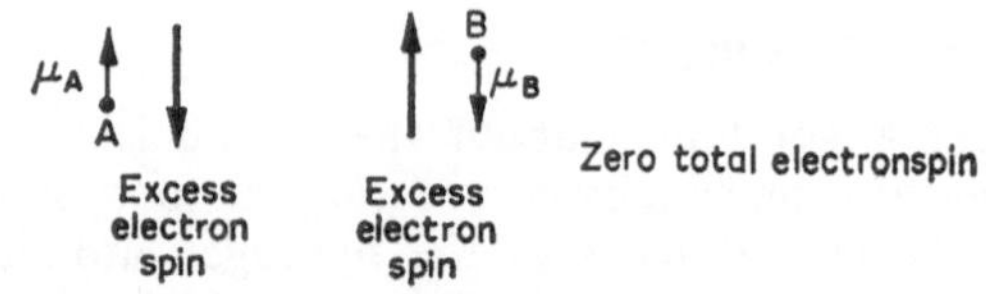

Figure 6-10.

This is a very useful result which we could obtain in another way, i.e. first we could consider just one of the nuclei which tends to unpair the electrons in its vicinity, thus producing an equal but opposite spin density at the other end of the bond, which then interacts with the second nucleus by the same mechanism as that responsible for hyperfine splitting in ESR.

In the case of the proton spin–spin coupling in H_2 the most favourable state of affairs is when the nuclear spins are anti-parallel, or in slightly different terms, if α-spin is preferred near one of the nuclei, then the tendency is for that nucleus to induce β-spin density in the vicinity of the other. The sign of the coupling constant is then defined to be positive. Negative coupling constants occur when in the state of lowest energy the nuclear spins are parallel.

The Averaged Energy Approximation

As we have already pointed out, the advantage of this approach is that we do not have to know the forms of excited-state wavefunctions. We have to be able, however, to make a reasonable estimate of the "averaged" energy ΔE and this is one of the weaknesses of the method. Be that as it may, the formula [see Eq. (6.5)] for the coupling energy of two nuclei, A and B reduces to the following:

$$hJ_{AB} = -4K_{AB} \sum_{i<j}^{N} \sum \langle 0 \mid \delta_i(A)\, S_{iz} \delta_j(B)\, S_{jz} \mid 0 \rangle\, I_{Az} I_{Bz} / \Delta E \tag{6.8}$$

where K_{AB} is a constant for the two nuclei, i.e.

$$(16\pi\beta_e\hbar/3)^2\, \gamma_A \gamma_B$$

When two nuclei are joined by what can be considered as an electron-pair bond the coupling constant should be positive, according to this formula, because the spin part of the bond wavefunction is $(\alpha\beta - \beta\alpha)/\sqrt{2}$, so that

$$S_{1z}S_{2z}(\alpha\beta - \beta\alpha)/\sqrt{2} = -\tfrac{1}{4}(\alpha\beta - \beta\alpha)/\sqrt{2}$$

i.e. $J_{AB} > 0$.

In fact, if we use simple MO theory in conjunction with the averaged energy approximation, the predicted coupling constants are always positive. Without going into too much detail we can see why this is so in the following way: the MO ground-state wavefunction is a single determinant $P^{-}(a\alpha a\beta b\alpha b\beta \ldots)/\sqrt{N!}$ where $a, b, \ldots$ are the MO's and P^{-} a unitary anti-symmetrizing operator. The individual integrals arising from Eqn. (6.8) are of types:

(i)

$$\langle(a\alpha a\beta b\alpha b\beta \ldots) \mid \delta_1(A)\, \delta_2(B)\, S_{1z}S_{2z} \mid (a\alpha a\beta b\alpha b\beta \ldots)\rangle$$

and (ii)

$$\langle(a\alpha a\beta b\alpha b\beta \ldots) \mid \delta_1(A)\, \delta_2(B)\, S_{1z}S_{2z} \mid (b\alpha a\beta a\alpha b\beta \ldots)\rangle$$

(i) will be negative and will occur in the expansion with a positive sign, whereas (ii) will be positive but occurs in the expansion with a negative sign because two electrons have been interchanged in going from one wavefunction in the integral to the other.

Other types of integral are either zero, due to the orthogonality of the MO's, or cancel each other; e.g. for each integral of the type $\langle(a\alpha a\beta b\alpha b\beta \ldots) \mid \ldots S_1S_{3\,z} \mid (a\alpha a\beta b\alpha b\beta \ldots)\rangle$ there is one which is equal to it but has the opposite sign, i.e.

$$\langle(a\alpha a\beta b\alpha b\beta \ldots) \mid \ldots S_{1z}S_{4z} \mid (a\alpha a\beta b\alpha b\beta \ldots)\rangle$$

Substitution into Eq. (6.8) leaves us with a positive coupling constant.

Comparison of MO and VB Calculations for H_2

It is not, of course, possible to observe the coupling energy of the protons in H_2 directly, since the protons are equivalent, so these brief calculations in the context of the averaged energy approximation are really for HD rather than H_2. If a and b are now to be the atomic orbitals associated with nuclei A, B respectively, the ground-state wavefunctions for the two molecular theories are:

$$\text{MO} = \tfrac{1}{2}(a+b)\,(a+b)\,(\alpha\beta-\beta\alpha)/\sqrt{2}$$

$$\text{VB} = (ab+ba)\,(\alpha\beta-\beta\alpha)/2$$

In units of $hK_{AB}/\Delta E$, the predicted coupling constants are, from Eq. (6.8)

(i) MO: $\tfrac{1}{4}\langle(a+b)\,(a+b) \mid \delta_1(A)\, \delta_2(B) \mid (a+b)\,(a+b)\rangle \approx \tfrac{1}{4}a_0^2b_0^2$

(assuming value of $a+b$ at nucleus A is $\approx a_0$, etc.)

(ii) VB: $\tfrac{1}{2}\langle(ab+ba) \mid \delta_1(A)\, \delta_1(B) \mid (ab+ba)\rangle \approx \tfrac{1}{2}a_0^2b_0^2$

Other things being equal it seems that the VB value for the coupling constant is twice that obtained from application of simple MO theory. The reason for this is that the ionic terms, which are not present in the simple VB function contribute nothing to the coupling because they tend to isolate the electrons on one of the atoms.

We can estimate the effects of electronegativity difference between the bonded atoms in the light of this, for this ionic terms become more and more important in the bond wavefunction as the electronegativity difference increases. Another consequence of a large difference in electronegativity difference between bonded atoms is that the bond excitation energy becomes large, so that we might expect that coupling constants should tend to be smaller when the nuclei are associated with orbitals which have widely different electronegativities.

6.7. VALENCE BOND THEORY OF PROTON COUPLING CONSTANTS

Now we come to the theory of coupling between nuclei which are not bonded directly together. In terms of the averaged energy approximation simple MO theory gives very poor results for geminal coupling constants, so we shall use only VB theory in this context. We simplify the calculations by considering a four orbital system which should give a reasonably good approach to geminal and vicinal coupling constants in hydrocarbons.

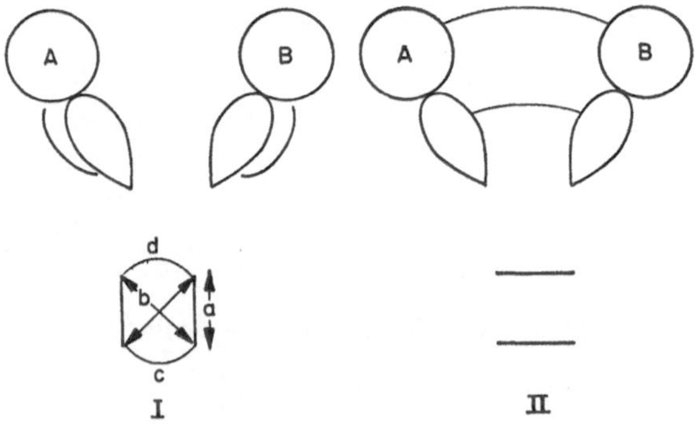

Figure 6-11. Canonical structures for four orbital system.

The problem is identical in form to that in section 6.4, the only differences being in the form of the perturbation and that the "phantom" orbital is replaced by one which interacts with the other atomic orbitals. The VB wavefunction is given by

$$|\,VB\rangle = c_{I}\,|\,I\rangle + c_{II}\,|\,II\rangle$$

and using the superposition diagrams as before we find

$$c_{II}/c_I = \tfrac{1}{2}(c+d-2b)/(2a-c-d) \tag{6.9}$$

In order to find the coupling constant, we now have to evaluate the integral

$$\langle VB \mid \mathscr{H}' \mid VB\rangle = c_I^2\langle I \mid \mathscr{H}' \mid I\rangle + 2c_Ic_{II}\langle I \mid \mathscr{H}' \mid II\rangle + c_{II}^2\langle II \mid \mathscr{H}' \mid II\rangle$$

as indicated by Eq. (6.8), and first we have to know the spin states represented by I and II, i.e.

for I $\quad \mid I\rangle = \mid \alpha\beta\alpha\beta - \alpha\beta\beta\alpha - \beta\alpha\alpha\beta + \beta\alpha\beta\alpha\rangle$

for II $\quad \mid II\rangle = \mid \alpha\beta\alpha\beta - \alpha\alpha\beta\beta - \beta\beta\alpha\alpha + \beta\alpha\beta\alpha\rangle$

Directly from the operation of $S_{iz}S_{jz}$ on these functions we find that

$$\langle I \mid \mathscr{H}' \mid I\rangle = \langle II \mid \mathscr{H}' \mid II\rangle = 0$$

and

$$\langle I \mid \mathscr{H}' \mid II\rangle \propto a_0^2 b_0^2$$

so that the coupling constant is proportional to

$$c_{II} \approx \tfrac{1}{2}(c+d-2b)/(2a-c-d)$$

Vicinal Coupling Constants—Connection with ESR

When the exchange integral c is much or appreciably larger than d or b, then the coupling constant is given by

$$J_{AB} = \text{constant}(c)/\Delta E$$

If now atom A, say, is removed and the rest of the system left unchanged then the coupling between the odd electron in the three orbital fragment and nucleus B is proportional to c also. This is an interesting connection between coupling constants in NMR and in ESR, and it implies for example, that vicinal coupling constants in NMR and β-splittings in ESR should vary in similar ways with dihedral angles. To be more explicit, in both cases the splitting should vary with c which can be written in the form $c_{\sigma\sigma} + c_{\sigma\pi}\cos\theta + c_{\pi\pi}\cos^2\theta$. The integral $c_{\pi\pi} = \int p_\pi(1)p_\pi'(2)p_\pi(2)p_\pi'(1)/r_{12}\,d\tau$ is much larger than the others here and this accounts for the observed variation of vicinal coupling in NMR, or of β-splittings in ESR. Since $c_{\pi\pi}$ is negative, being similar to the resonance integral responsible for the π-bond in ethylene, we can expect the coupling constants to be positive.

Geminal Coupling Constants

When the two bonds are to the same atom it is rather more difficult to make an accurate estimate of the relative magnitudes of the integrals involved. For methane the integrals which determine the sign of the coupling constant as well as its magnitude, after we have accepted that a is relatively large and negative, have been calculated to be $d = -1.0$eV; $b = +0.233$ eV; $c = +1.0$ eV; which means that the expression $(c+d-2b)$ is negative and that therefore the coupling constant is positive. The actual magnitude comes out to be 14 c/s, in exact agreement with the observed value! Unfortunately the sign of the observed splitting is negative.

It is possible to rationalize this apparent discrepancy between theory and experiment, for the integrals were estimated using an effective atomic number of hydrogen of unity. It seems probable, however, that the atomic orbital to use for hydrogen in a molecule should be more contracted than this implies, i.e. with $Z_{eff} = 1.2$. If the integration were carried out bearing this in mind, one would find that the "long-range" integrals b and d would be smaller than their values given above, so it is quite possible that the sign of the coupling constant would change. This discussion shows that we must take great care in placing significance on the results of calculations which depend on relatively small differences between large, uncertain quantities.

6.8. MOLECULAR ORBITAL THEORIES OF COUPLING CONSTANTS

The expression for coupling constants which is obtained by using simple MO theory and the averaged energy approximation is particularly concise. To make things clearer let ψ_r be the rth filled MO, so the MO wavefunction is

$$P^{-}(\psi_1\alpha\psi_1\beta\psi_2\alpha\psi_2\beta\ldots)/\sqrt{N!} \qquad \text{(for } N \text{ electrons)}$$

The integral in Eq. (6.8) becomes,

$$\left\langle \psi_1\alpha\psi_1\beta\psi_2\alpha\psi_2\beta\ldots \middle| \sum_{i<j}^{N}\sum \delta_i(A)\,\delta_j(B)\,S_{iz}S_{jz} \middle| P^{-}(\psi_1\alpha\psi_1\beta\psi_2\alpha\psi_2\beta\ldots)\right\rangle$$

As we said before, most of the terms in this expansion cancel out and it reduces to

$$-\tfrac{1}{4}\sum_r\sum_s \psi_{rA}\psi_{sB}\psi_{sA}\psi_{rB}$$

where ψ_{rA} is, for example, the value of MO ψ_r at nucleus A. The expression for the coupling between nuclei A and B then becomes:

$$hJ_{AB} = +K_{AB}(\Sigma_r\psi_{rA}\psi_{rB})^2/\Delta E \qquad (6.10)$$

Calculation of Coupling Constants using MO Excited States

As we have already indicated, another way of trying to estimate coupling constants is to ignore most of the higher excited states and include only those which are relatively easy to obtain in an approximate form. The most obvious general theory to use here is simple MO theory, and the results obtained seem to be an improvement over those obtained from MO theory plus the averaged energy approximation, for it is now possible to predict negative coupling constants. The general formula obtained is nearly as simple as that in Eq. (6.10), and we can derive it as follows:

Let ψ_r be the rth occupied MO and ϕ_s be the sth empty MO, in the ground-state of a molecule. The antisymmetrized wavefunction for the ground-state will be,

$$|\,0\rangle = P^{-}|\,[\psi_1\psi_1(\alpha\beta-\beta\alpha)/\sqrt{2}]\;\psi_2\alpha\psi_2\beta\ldots\rangle/(N!/2)^{1/2}$$

and the appropriate triplet-state function arising, for example, from an excitation from ψ_1 to ϕ_s will be

$$|\,1, s\rangle = P^{-}|\,[\psi_1\phi_s-\phi_s\psi_1)/\sqrt{2}(\alpha\beta+\beta\alpha)/\sqrt{2}]\;\psi_2\alpha\psi_2\beta\ldots\rangle/(N!/2)^{1/2}$$

In both of these formulae we have already made the function antisymmetrical with respect to the parts involving ψ_1 and ϕ_s, for clarity. P^{-} has to make the whole function antisymmetrical. The perturbations we require are $\mathscr{H}_A' = \sum_i \delta_i(A)\, S_{iz}K_A$; and $\mathscr{H}_B'$; and the corresponding matrix elements are, for example, in the above case:

$$\langle 0\,|\,\mathscr{H}_A'\,|\,1\rangle = \Big\langle P^{-}\left[\psi_1\psi_1\left(\frac{\alpha\beta-\beta\alpha}{\sqrt{2}}\right)\psi_2\alpha\psi_2\beta\ldots\right]|\,\mathscr{H}_A'\,|$$

$$\times\left[\frac{(\phi_s\psi_1+\psi_1\phi_s)}{\sqrt{2}}\frac{(\alpha\beta+\beta\alpha)}{\sqrt{2}}\right]\psi_2\alpha\psi_2\beta\ldots\Big\rangle$$

$$= K_A\Big\langle\psi_1\psi_1\frac{(\alpha\beta-\beta\alpha)}{\sqrt{2}}\,|\,\delta_1(A)\,S_{1z}+\delta_2(A)\,S_{2z}\,|$$

$$\times\frac{(\psi_1\phi_s-\phi_s\psi_1)}{\sqrt{2}}\frac{(\alpha\beta+\beta\alpha)}{\sqrt{2}}\Big\rangle$$

$$\text{i.e.}\quad \langle 0\,|\,\mathscr{H}'\,|\,1, s\rangle = -K_A\psi_{1A}\phi_{sA}/\sqrt{2}.$$

Analogous formulae will apply for all the other possible single electron excitations, so from Eq. (6.4), the coupling energy between A and B will be

$$hJ_{AB} = K_{AB}\sum_{\substack{r\\ \text{occupied MO's}}}\;\sum_{\substack{s\\ \text{empty MO'}}}\psi_{rA}\psi_{rB}\phi_{sA}\phi_{sB}/(E_s-E_r) \qquad (6.11)$$

Usually our MO's are found in terms of the atomic orbitals, i.e. we use the LCAO method, and in that case Eqs. (6.10) and (6.11) may be written in terms of the coefficients of the AO's associated with *A* and *B*:

$$hJ_{AB}=k_{AB}\left(\sum_r c_{rA}c_{rB}\right)^2\Big/\Delta E \tag{6.10a}$$

$$hJ_{AB}=k_{AB}\sum_r\sum_s c_{rA}c_{rB}c_{sA}c_{sB}/(E_s-E_r) \tag{6.11a}$$

where k_{AB} is a constant for the two nuclei and c_{rA}, for example, is the coefficient of an orbital associated with atom *A* in the MO ψ_r, etc. If there is more than one type of AO associated with *A*, or for that matter under any circumstances, c_{rA} is the coefficient of the *s*-type orbital of *A*, in the MO ψ_r.

Example: coupling in molecule AB_n where all of the *B*'s are equivalent.

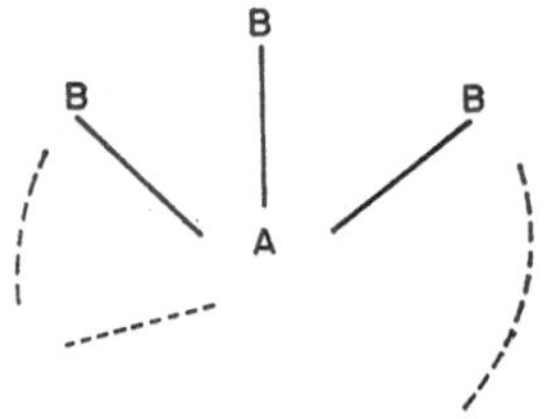

Figure 6-12.

Only two MO's will involve the *A* *s*-orbital and they will be totally symmetric. Ignoring differences in electronegativity, they are:

$$\text{bonding} \quad (\psi_A+\Sigma\psi_B/\sqrt{n})/\sqrt{2}$$

$$\text{antibonding} \quad (\psi_A-\Sigma\psi_B/\sqrt{n})/\sqrt{2}$$

Both the averaged energy expression Eq. (6.10a) and the other Eq. 6(.11) lead to the same type of formula for the coupling energy, i.e.

$$hJ_{AB}=k_{AB}/4n\Delta E$$

though ΔE means something slightly different in the two cases. However, both approaches lead us to expect that the coupling between *A* and *B*, which are neighbouring nuclei, should be roughly inversely proportional to the number of neighbours of *A* when differences in electronegativities are small. This certainly seems to be true for hydrocarbons as we can see in table 6.II, suggesting that the excitation energies must be similar in these compounds. The ^{13}C–^{13}C coupling constants in these molecules also follow the expected trend, i.e. since

Table 6.II ^{13}C-H coupling constants in simple hydrocarbons

Molecule	Number of neighbours to carbon	$J_{^{13}C-H}$ (Hz)	$n \times J$
Ethane	4	124.9	500
Ethylene	3	156.4	469
Acetylene	2	248.7	497

in these cases both atoms have n neighbours the splittings should be inversely proportional to n^2, and this is approximately so, though the proportionality constant is not simply related to that applicable to the above ^{13}C–H couplings.

Example: Vicinal coupling constants

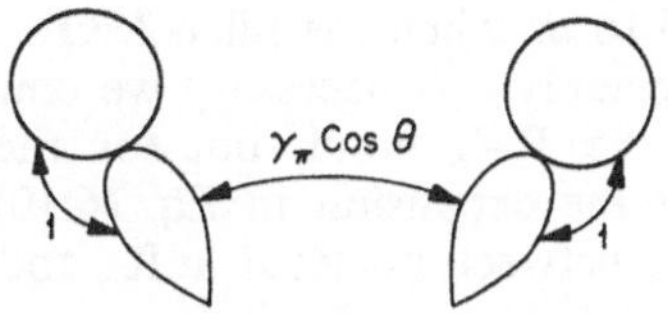

Figure 6-13.

If as before we look at a four orbital system consisting of two bonds which interact only via one pair of orbitals (see Figure 6–13), it is not difficult to obtain the MO's and thence the coupling constants. In this highly simplified example, the energy levels are given by

$$(\alpha - E) = (\pm z \pm \sqrt{4+z^2})/2 \text{ where } z = \gamma_\pi \cos\theta.$$

The coefficients of these 4 MO's are easily found and then fed into Eqs. (6.10) and (6.11), both of which lead to the result that J_{AB} should vary with $\cos^2\theta$; θ being, as before, the dihedral angle.

Non-bonding Orbital Approach

One trouble with the MO approach is that whereas the molecular orbitals are usually delocalized, the interactions between the electron spins with a nucleus are rather strongly localized, i.e. near to that nucleus. The result of this is that more or less all the terms in the expansions Eqs. (6.10) and (6.11) have to be included. It would be very useful if it could be arranged so that most of the terms could be neglected, for example, by using appropriate equivalent orbitals. A more simple approach is to consider situations in which a molecule may be considered as being made up of fragments which interact to form further bonds which hold the complete molecule together; e.g. ethane may be regarded as being composed of two "tetrahedral" CH_3 groups which interact largely via the fourth sp^3 hybrid orbitals of each carbon atom, or equally as a hydrogen atom + an ethyl fragment. The interactions between the two fragments are regarded as perturbations and the method is often used to give a quick estimate of delocalization energies in aromatic π-electron systems. We shall make use of it in the following way:

Suppose the fragments R– and S– have each one electron in a non-bonding orbital (NBO), then to the first-order, the bond between the two fragments arises from the interaction of these two orbitals, and the MO ground-state wavefunction for the molecule will be approximately given by assigning pairs of electrons to the bonding MO's localized in one fragment or the other, and one pair to the bonding orbital which is a linear combination of the two NBO's. This approximation is only useful to us when the filled MO's in the two fragments do not interact appreciably. If necessary we could consider a "localized" excitation in the R–S bond, but for the moment it is only necessary to see that the expansion in Eq. (6.10) reduces to a single term, and the coupling between nuclei A in R– and B in S– is given by:

$$hJ_{AB} = k_{AB} c_A^2 c_B^2 / 4\Delta E \tag{6.12}$$

where, for example, c_A is the coefficient of the AO of A in the NBO of R—, or rather the coefficient of the s orbital associated with A in the NBO of R–.

In the case when one of the fragments is a proton, the coupling with the other nucleus will depend only on the square of its AO coefficient in the other fragment which also determines its splitting of the ESR spectrum of the radical fragment. This points towards an even more complete connection between ESR and NMR than that already indicated between vicinal coupling constants in RH and β-splittings in radical R–. Some empirical comparisons are given in Figure 6-14.

This approach also gives us a very good picture as to how nuclear spin–spin coupling constants arise, i.e. a magnetic nucleus polarizes the electron spins in its vicinity. Electron spin density in more remote parts of the molecule must therefore appear having an overall sign opposite to that induced near to the nucleus in order to keep the total electron spin of the molecule equal to zero. Thus if the proton A in R–H_A induces an excess spin of $\delta\alpha$ in its atomic orbital, then the fragment R– will be left with an excess spin of $\delta\beta$.

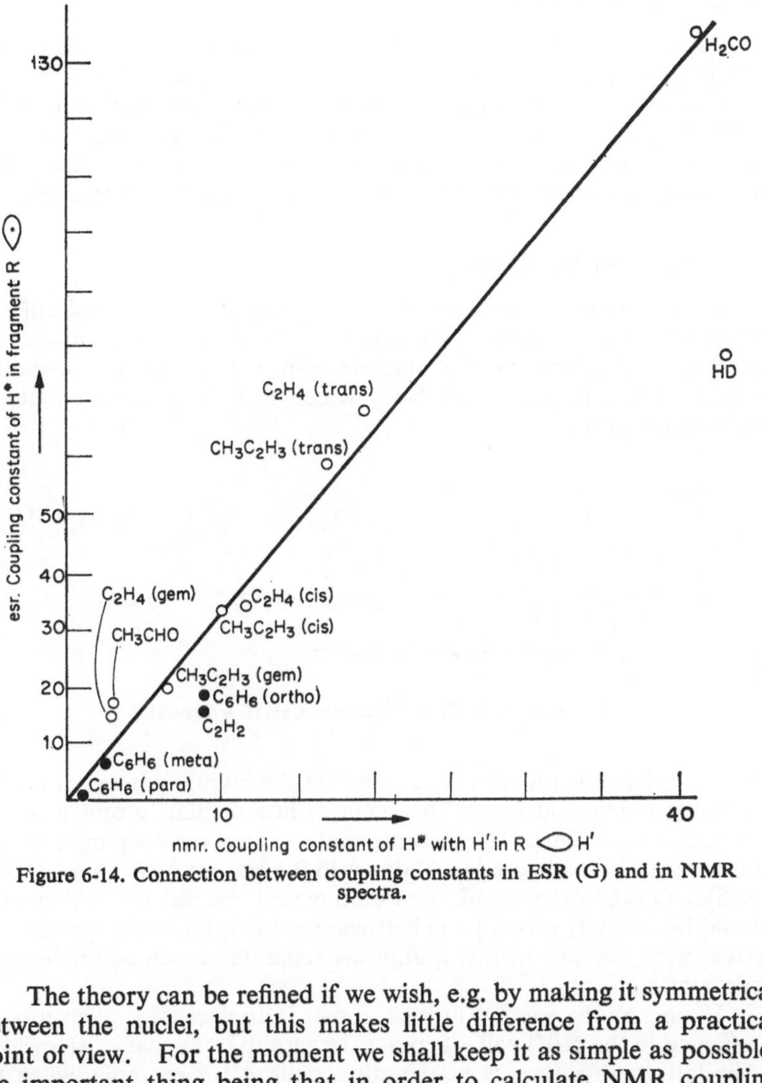

Figure 6-14. Connection between coupling constants in ESR (G) and in NMR spectra.

The theory can be refined if we wish, e.g. by making it symmetrical between the nuclei, but this makes little difference from a practical point of view. For the moment we shall keep it as simple as possible, the important thing being that in order to calculate NMR coupling constants all we have to do is to find the AO coefficients of a non-bonding orbital.

The excitation energy ΔE is generally to be taken as an empirical parameter and for proton–proton coupling it seems to be a reasonable approximation, at least in hydrocarbons, simply to absorb it into the constant k_{AB}.

E

Vicinal Coupling Constants

These are calculated from the corresponding fragments, for example, in ethylene, they are determined by the coupling constants in the vinyl radical, or rather, in theoretical terms, the coefficients of the NBO of vinyl, which were calculated in section 6.3 (see Figures 6-1(a) and (b)). Also the same dependence on dihedral angle is obtained as those predicted from the various other theories we have discussed.

Long-range Coupling Constants

Let us consider as an example of long-range coupling, molecules of the type $CH_3(C:C)_nCH_3$. We can estimate the coupling between the protons by looking at the odd-electron orbital in the σ-radical $-CH_2(C:C)_nCH_3$ (see Figure 6-15) which is a NBO if we equate all coulomb integrals.

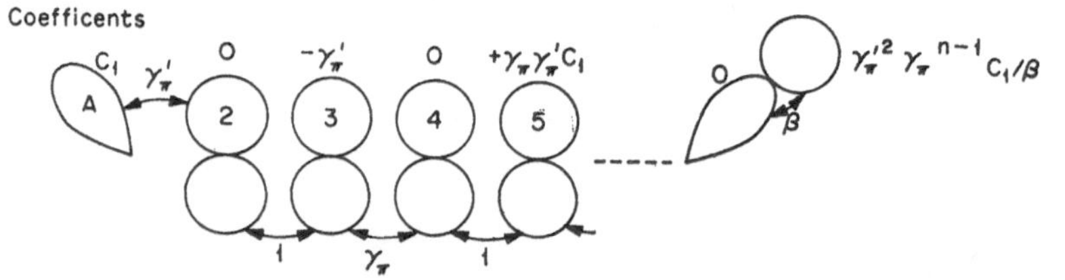

Figure 6-15. NBO coefficients in $CH_3(C{=}C)_n CH_3$.

Since J_{AB} depends on c_B^2 we can see that the coupling between methyl protons gets less and less as the chain of intermediate atoms increases in length, only because the bonds alternate. Thus the coupling constant falls from 8 Hz in ethane ($n=0$) down to 0.4 Hz in $CH_3(C:C)_3CH_2OH$ ($n=3$). On the other hand, we would expect, on this basis, that there should be virtually no coupling between methyl protons via a π-electron system when the two methyl groups are separated by an odd number of carbon atoms.

If one of the methyl groups above is replaced by a proton, its coefficient in the NBO will, of course, be zero, by symmetry. However, a similar situation occurs in the ethyl radical, for example, when the two types of proton have splittings of the same order of magnitude but of opposite signs. Using the same idea, i.e. McConnell's relation Eq. (6.3), we are not too surprised to find that the coupling between the protons at the ends of the molecule $CH_3(C:C)_3H$, is of the same order as that of the above compound (0.65 Hz); but we would expect the sign to be different.

The example we have just considered involves configuration interaction at one end of the chain in order to get "partial spin density" on to one of the protons. By the same token coupling can occur between protons via a π-electron system even when they are both in its nodal plane. In this case spin polarization occurs at both ends of the chain but delocalization of the effects takes place in the π-electron orbitals. (See Figure 6-16.)

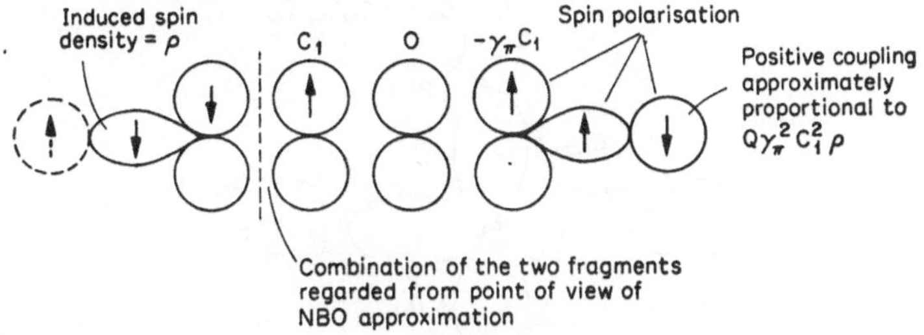

Figure 6-16.

Evidently this approach is very useful because it enables more complicated situations to be tackled without excessive computation, compared to the standard MO approach.

As another example let us look at the problem of proton–proton splitting in simple (planar) cyclic systems. These molecules may be neutral, positively or negatively charged. The vicinal coupling constants will vary with the ring size, for in all the theories we have applied to the problem, J_{vic} varies with γ_π. But γ_π will increase with ring size,

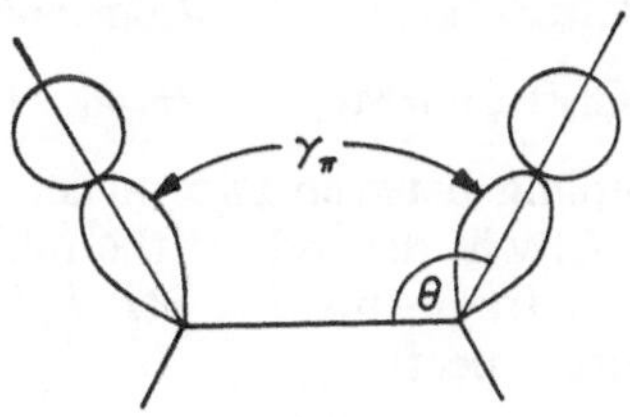

Figure 6-17.

since the hybrid orbitals in the directions of the C–H bonds become more and more nearly perpendicular to each other as n increases (i.e. θ decreases). This makes their π-overlap more efficient and hence should increase γ_π and hence the vicinal coupling constant. This is the trend which is observed, e.g. the coupling constants for the ethylenic protons in cyclohexene is 9.4 Hz; in cyclopentene is 5.1 Hz; in cyclobutene is 2.8 Hz, and in cyclopropene is less than 2 Hζ.

As a final example of this approach let us consider the coupling constants in certain complexes of the platinum group of metals.

We choose as a model, an appropriate set of atomic orbitals such that the only significant interaction between the metal and a ligand is through a two electron bond, formed from a hybrid metal orbital (*dsp*^{2} for square planar and d^2sp^3 for octahedral complexes) and a single MO of the ligand.

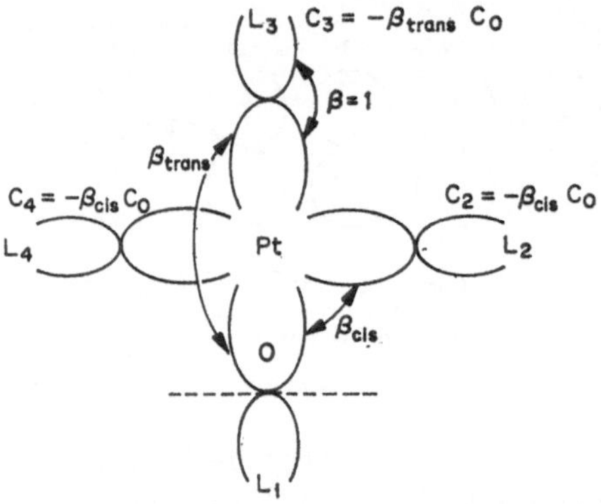

Figure 6-18. Square planar system with NBO coefficients in fragment.

First, the coupling between a nucleus in a ligand and the metal nucleus will depend mainly on the amount of "*s*" character of the metal orbital in the bond. The coupling constants in 4- and in 6-coordinate compounds should be there in the ratio $\sim\frac{1}{4}/\frac{1}{6}$, i.e. $\sim\frac{3}{2}$. This agrees with observation, e.g.

trans $PtCl_2(PEt_3)_2$ $\quad J_{P-Pt} \sim 2800$ Hz

trans $PtCl_2(PEt_3)_2Me_2$ $\quad J_{P-Pt} \sim 1900$ Hz

Secondly, the coupling between two nuclei one in one ligand L_1, the other in L_2, or in L_3 will depend on the odd electron densities at these nuclei in the two fragments L_1, $PtL_2L_3L_4$ (as before). From Figure 6-18 we therefore expect

$$\frac{J_{cis}}{J_{trans}} = \frac{c_{cis}^2}{c_{trans}^2} = \frac{\beta_{cis}^2}{\beta_{trans}^2}$$

But

$$\beta_{cis} = \tfrac{1}{4}(\alpha_s - \alpha_d) \qquad \text{square planar}$$

$$\text{or } \tfrac{1}{6}(\alpha_s - \alpha_d) \qquad \text{octahedral}$$

and

$$\beta_{trans} = \tfrac{1}{4}(\alpha_s + \alpha_d - 2\alpha_p) \qquad \text{square planar}$$

$$\text{or } \tfrac{1}{6}(\alpha_s + 2\alpha_d - 3\alpha_p) \qquad \text{octahedral}$$

and since in these metals, the "*p*" orbital involved starts by being empty and is of much higher energy than the *s* or *d* orbitals (which have comparable energies), the coulomb integrals are in the order $|\alpha_s| \sim |\alpha_d| \gg |\alpha_p|$ and hence we expect $J_{cis} \gg J_{trans}$. This is in agreement with the coupling of ^{31}P nuclei which are of the order of 150 Hz when they are trans- but are much smaller (~15 Hz) when they are cis- with respect to each other.

Chapter 7

Time-dependent Effects in Magnetic Resonance

7.1. CONSEQUENCES OF BROWNIAN MOTION

We have already come across the fact that the trace of a magnetic resonance spectrum consists of a series of peaks each having some more or less well-defined shape. Even where there is only one line, it is clear that absorption takes place over a range of frequencies for a given applied field, and this can be ascribed to the fact that the field acting on an individual spin is modified by its immediate environment, so that different spins experience different instantaneous fields. In solutions, these local fields are fluctuating rapidly due to Brownian movement, i.e. molecules in the liquid are vibrating, rotating and moving about rapidly and at random. In the following discussion we shall use the word "spin" for the moment of the particular nuclear-type or odd electron whose resonance is being observed.

Over a long period, the average fields at the spins will all be the same; whereas over a short period, the time-average field will differ from one spin to another. There are two extreme situations, one in an extremely viscous medium, such as a glass or a solid, where the fluctuations are relatively slow so that the resonance is broad due to the large variations in local average fields; and another in a highly mobile liquid, where the molecules tumble about so violently that the spins are in the same average field. In this case the resonance will be sharp and the term "motional narrowing" is used to describe what is happening.

It is a useful exercise, at this point, to be a little more quantitative and use the characteristics of Brownian motion to derive the Bloch equations, which we shall use, in one form or another, in examining relaxation effects.

Calculations of T_1

Suppose the fluctuating part of the field at a spin is B'. Then it is the x- and y-components of B' which will tend to induce transitions between the two energy levels α and β, for particles spin $\frac{1}{2}$, which cause relaxation of the z-component of the total spin of the sample, i.e. according to the results of sections 3.4 and 3.6.

Before proceeding further it is worth noting that this fluctuating field B' has several characteristics in common with random fluctuations in general, and these are:

(i) The mean value of B' is zero, since the field is random and is just as likely to be in one direction as in any other. The mean here can be taken over either a large number of spins or a single spin observed over a sufficiently long period of time.

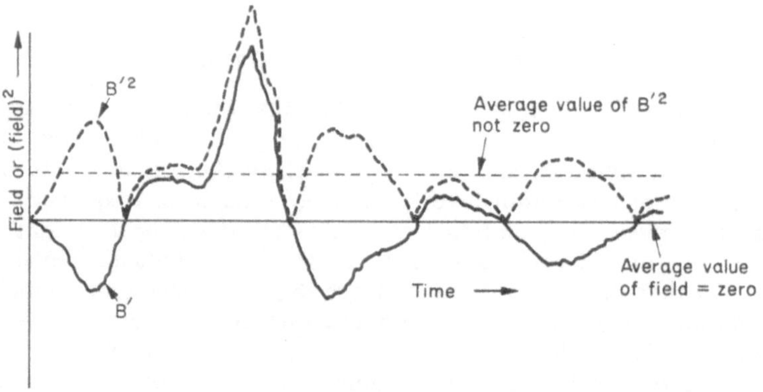

Figue 7-1. Random fluctuations of field.

(ii) The mean value of B'^2 will not, in general, be zero but will have some definite value, since this function of the field is always positive whatever the direction of the field. For a large number of spins the mean value of B'^2 will be independent of the time, i.e. when some sort of steady-state conditions prevail.

(iii) The mean value of $B_x'(t)B_x'(t+\tau)$ will be zero if τ is large, because after a sufficiently long period of time $B'(t+\tau)$ will be as likely to be in any one direction as any other, whatever its direction was at time t. For large τ: $\overline{B_x'(t)B_x'(t+\tau)}=\overline{B_x'(t)}.\ \overline{B_x'(t+\tau)}=0$

Another way of saying this is that for a large time interval, τ, the average field at $t+\tau$ does not depend on that at time t, i.e. the two fields are not correlated.

When τ is small enough, however, $B'(t+\tau)$ may not be very different from $B'(t)$, so that the average of the product $B_x'(t)\,B_x'(t+\tau)$ will not then be zero, and the fields at the two times are more or less correlated. This function $k(\tau)=\overline{B'(t)\,B'(t+\tau)}$ is called the correlation function for the field B' and so far we have been deducing some of its properties from "common-sense".

Some other useful properties are, first that since average values should not depend on the time, when steady-state conditions exist,

$$k(\tau)=k(-\tau)$$

and from the identity:

$$[B'(t)\pm B'(t+\tau)]^2=B'^2(t)+B'^2(t+\tau)\pm 2B'(t)B'(t+\tau)$$

$$k(\tau)<\overline{B^2(t)}$$

The function which we shall use and which satisfies all our intuitive requirements is of the type:

$$k(\tau)=\overline{B'^2}\exp\left(-\mid\tau\mid/\tau_c\right) \qquad (7.1)$$

We shall also assume that the different components of B' are not correlated with each other, i.e. $\overline{B_x'(t)\,B_y'(t)}=0$, etc., but that they do all share the same correlation time τ_c.

Now let us return to the calculation of T_1. Under the influence of a time-dependent perturbation $\mathscr{H}'$ we can write the wave function of a system which can exist in two states $\mid a\rangle$ and $\mid b\rangle$, as a linear combination, i.e. $\mid t\rangle=c_a\mid a, t\rangle+c_b\mid b, t\rangle$. The coefficients are given by the "key" equations of time-dependent perturbation theory, e.g.

$$i\hbar\partial c_b/\partial t=c_a\langle b\mid\mathscr{H}'\mid a\rangle+c_b\langle b\mid\mathscr{H}'\mid b\rangle$$

Now if the system starts in state $\mid a\rangle$ ($c_a=1$, $c_b=0$) then the probability per unit time that a transition will occur is $(\partial/\partial t)\mid c_b\mid^2$. We know $\partial c_b/\partial t$ already and we obtain c_b from it by integration, i.e.:

$$i\hbar c_b=\int_0^t\langle b\mid\mathscr{H}'(\xi)\mid a\rangle\exp(i\omega\xi)\,d\xi,\ \text{where}\ \omega=(E_b-E_a)/\hbar$$

The probability per unit time, P_{ab}, becomes, after a little manipulation

$$\hbar^2P_{ab}=2\int_0^t\langle a\mid\mathscr{H}'(t)\mid b\rangle.\langle b\mid\mathscr{H}'(\xi)\mid a\rangle\cos\omega(t-\xi)\,d\xi \qquad (7.2)$$

This is a general formula; in our present case the fluctuating part of the Hamiltonian, $\mathscr{H}'(t)$, is $-\bar{\mu}.\bar{B}'(t)$, so that after evaluating the matrix elements and taking an average over all the spins and using

Eq. (7.1), the transition probability/unit time becomes:

$$P_{ab}=\tfrac{1}{2}\gamma^2\int_0^t \overline{(B'^2_x+B'^2_y)}(\exp(-\mid t-\xi\mid/\tau_c))\cos\omega(t-\xi)\,d\xi$$

i.e.

$$P_{ab}=\tfrac{1}{2}\gamma^2\overline{(B'^2_x+B'^2_y)}\int_0^t \exp(-\tau/\tau_c)\cos\omega\tau\,d\tau\quad\left\{\begin{array}{l} t\geqslant\tau\geqslant 0\\ \tau=t-\xi\end{array}\right. \tag{7.3}$$

Finally we obtain the expression we require for T_1 in terms of the correlation time:

$$1/T_1=2P_{ab}=\gamma^2(B_x'^2+B_y'^2)\,\tau_c/(1+\omega^2\tau_c^2) \tag{7.4}$$

Calculation of T_2

Having obtained an expression for T_1 using quantum mechanics it is interesting to approach the parameter T_2 from a classical "resonance" viewpoint. We start from the equation of motion for a spin:

$$d\bar{\mu}/dt=\bar{\mu}\wedge\bar{B}$$

Next we transform to a rotating coordinate system (XYZ) defined by the equations which follow, i.e. in terms of the stationary system, xyz, and the angular velocity $\omega=\gamma B_0$, where B_0 is the applied field in the z-direction.

$$m_Z=\mu_z \qquad b_Z=B_z-B_0=B_z'$$

$$m_X=\mu_x\cos\omega t-\mu_y\sin\omega t \qquad b_X=B_x'\cos\omega t-B_y'\sin\omega t$$

$$m_Y=\mu_y\cos\omega t+\mu_x\sin\omega t \qquad b_Y=B_y'\cos\omega t+B_x'\sin\omega t$$

In the rotating frame the equation of motion is very simply:

$$d\bar{m}/dt=\gamma\bar{m}\wedge\bar{b}$$

For the X-component we find:

$$(1/\gamma^2)\,dm_X/dt=b_Z\int_0^t(m_Zb_X-m_Xb_Z)\,d\xi\,.\,b_Y\int_0^t(m_Xb_Y-m_Yb_X)\,d\xi$$

Leaving out the terms which go to zero when averaged, this simplifies to

$$dm_X/dt=-\gamma^2\int_0^t[b_Z(t)\,b_Z(\xi)+b_Y(t)\,b_Y(\xi)]\,m_X(\xi)\,d\xi$$

Now if m_X varies only slowly relatively to the fluctuations in the local fields we can compare this with the Bloch equations and get the appropriate expression for $1/T_2$. Thus, changing back to the stationary coordinate system, we obtain:

$$1/T_2=\gamma^2\int_0^t \overline{(B_z'(t)\,B_z'(\xi)+B_y'(t)\,B_y'(\xi)}\cos\omega(t-\xi))\,d\xi$$

Again using the correlation function of Eq. (7.1) and using the symmetry between the x- and y-directions we finally obtain:

$$1/T_2=\gamma^2 B_z'^2\tau_c+1/2T_1 \tag{7.5}$$

In deriving Eqs. (7.1) and (7.5) we have assumed that the time t is small relative to the time scale of our experiment but large relative to τ_c.

In solutions average values are generally independent of direction, so that $B_x'^2=B_y'^2=B_z'^2$, so if $\omega\tau_c$ is small $T_1=T_2$. However, as τ_c gets larger, T_1^{-1} eventually becomes small whereas $1/T_2$ gets larger. Figure 7-2 illustrates the type of dependence of the relaxation times on the correlation time.

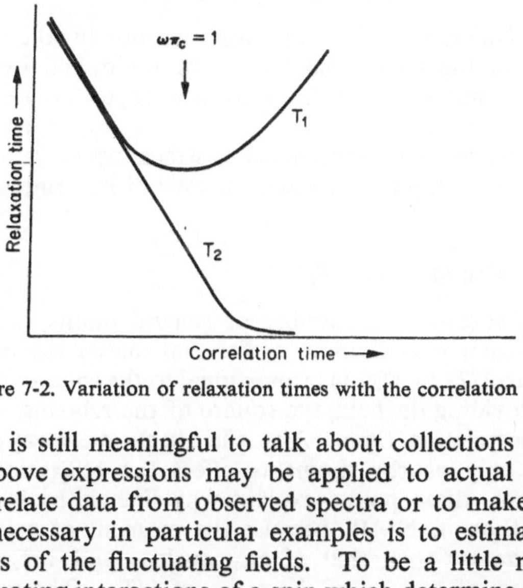

Figure 7-2. Variation of relaxation times with the correlation time.

While it is still meaningful to talk about collections of individual spins, the above expressions may be applied to actual situations in order to correlate data from observed spectra or to make predictions. All that is necessary in particular examples is to estimate the mean square values of the fluctuating fields. To be a little more precise, it is the fluctuating interactions of a spin which determine its relaxation times. However we can usually treat these interactions as arising from effective fields, which can be due to a wide variety of causes. Fortunately, the effective fields from different sources contribute independently to $1/T_1$ and to $1/T_2$, because of the random nature of these fields.

Example: Intramolecular relaxation in water: At ordinary temperatures and at the fields used in NMR, it is a good approximation to equate T_1 with T_2, for $\tau_c \sim 10^{-12}$, $\omega \sim 10^8$. By τ_c here, we mean the rotational correlation time, which is also applicable to measurements of dielectric relaxation in this medium. From either of the Eqs. (7.1) or (7.5) we find:

$$(1/T_1)_{\text{intra}} = \tfrac{2}{3}\gamma^2 B'^2 \tau_c$$

To find the average value of the squared field B'^2 we remember that it arises from a dipole distance r away, making an angle with the line joining the two nuclei, so B' resolves into two parts, one radial, i.e. along the line of the nuclei, equal to $2\mu \cos \theta/r^3$; and the other transverse $= \mu \sin \theta/r^3$. The total average field is therefore,

$$\overline{B'^2} = \langle \mu^2(4 \cos^2 \theta + \sin^2 \theta)/r^6 \rangle$$

but $\mu = \gamma'\hbar I$ and $\langle I^2 \rangle = i(i+1)$; and $\langle \cos^2 \theta \rangle = \frac{1}{3}$, hence

$$(1/T_1)_{\text{intra}} = \gamma^2 \gamma'^2 \hbar^2 \tau_c / r^6 \qquad (7.6)$$

When the equivalence of the two protons is properly taken into account the formula on the right-hand has to be multiplied by 3/2, on the other hand, it might well apply in the form given, to HF, or in a modified form to HD or HOD.

The agreement with experiment is exact, which is surprising for such a "non-ideal" liquid like water, in view of its complexity.

Further Discussion of T_1 and T_2

Eq. (7.6) illustrates a number of general points, which are that the intramolecular contribution, or for that matter the intermolecular contribution to $1/T_1$ or $1/T_2$ is proportional to the square of the moment of the spin providing the field, the square of the relaxing spin moment, and to the correlation time. Also the effects die away rapidly with distance, so only nearby moments affect the relaxation of a spin. Taken together, these points explain why linewidths in ESR are so much larger than in NMR, because the moment of an odd electron is so much greater than that of any nucleus and $1/T_2 \propto (\text{moment})^2$. Linewidths are, of course, of the order of $1/T_2$. These relationships also explain why addition of paramagnetic ions broaden NMR lines of a sample and why the lines get broader the greater the concentration of magnetic species (for then the average distance is smaller, i.e. $1/r^6$

is larger and hence $1/T_2$ is also larger). This can be investigated in a controlled way by adding known amounts of paramagnetic transition metal ions to the solution.

The dependence on τ_c can be investigated by varying the viscosity of the medium, say by changing the temperature or by dissolution in a suitably viscous medium. High viscosity slows down the tumbling of the molecules and thereby increases the correlation time; linewidths therefore tend to increase with increasing viscosity, other things being equal.

The effective field acting on a spin can change due to factors other than its dipole–dipole interaction with other spins. An obvious example is when the *g*-tensor or the chemical shift are anisotropic, for then tumbling will produce a time-varying factor into the Hamiltonian. Thus in ESR, whenever the conditions in a radical lead to anisotropy of the *g*-tensor, i.e. when there are low-lying excited states and appreciable spin–orbit coupling, the relaxation is so efficient that that the signals are too broad to be observed. Similarly, when the spin of a radical is greater than $\frac{1}{2}$, the anisotropic dipolar interactions are usually so strong that such molecules give rise to no observable ESR spectrum.

The ideal conditions for observing ESR spectra in solution are therefore when there is little anisotropy of the *g*-tensor, such as in organic radicals, which ensures minimum intramolecular contribution to $1/T_2$; and also when the solutions are dilute, to ensure minimum intermolecular contribution. These conditions are found in dilute solutions of free radicals and transition metal ions in which the orbital angular momentum is effectively quenched.

When hyperfine coupling constants are anisotropic they will contribute to the intramolecular relaxation when the radicals are tumbling in solution. However different nuclear spin states will correspond to different effective fields acting on the odd electron, and will therefore lead to different relaxation times. This is reflected in the linewidths, peak heights and in the resolution of the lines making up the ESR spectrum.

The last case we shall deal with in this section is an effect which is sometimes observed in ESR spectra, which is that signals often become sharper as solutions are made more concentrated. This phenomenon, which is connected with the unusually sharp lines in the ESR spectra of crystals such as DPPH, is unexpected in view of the principles we have been discussing. What is happening here is that there is an interaction between the spins on neighbouring radicals of the type $A_{12}\bar{S}_1.\bar{S}_2$, which is modulated by the relative movement (vibrations etc.). This coupling will mix the states $\alpha\beta$ and $\beta\alpha$ and the time-dependency means that if initially there is an α-spin on radical *A* and β-spin on radical B, then after a time there will be a finite probability that the spins will have interchanged. The effect of this on, say, the α-spin is that its anisotropic interactions will change, and when the process takes place often enough, they will be averaged out and the resonance will become sharp. The phenomenon is called "exchange narrowing".

7.2. NUCLEAR QUADRUPOLE RELAXATION

Certain rather strange features appear in the proton resonance spectra of amines and of halogen compounds, for there is never any sign of, for example, splitting due to ^{35}Cl or ^{37}Cl nuclei, in spite of the fact that they both have spins of 3/2. On the other hand, proton resonance spectra of nitrogen compounds are often temperature-dependent, thus in pyrrole, for example, the triplet (1, 1, 1) splitting from the nitrogen, of the resonance spectrum of the nearby proton, appears only at relatively high temperatures, then as the temperature is lowered, the three lines broaden and finally reappear as a singlet at about $-40\,°C$. These effects can be explained in terms of the quadrupole moments of these nuclei, which have spins greater than $\frac{1}{2}$.

In the first place a quadrupole experiences a torque whenever it is in an inhomogeneous electric field, and therefore tends to be aligned along the line of maximum electric field gradient. However, the nuclei are held at certain orientations to the applied magnetic field, so if the time-varying torques acting on the quadrupole are strong enough, they may cause the nuclear moment to flip over to one of its other possible orientations, i.e. transitions between the magnetic levels can be induced by fluctuating electric field gradients acting on the nuclear quadrupole.

For this relaxation mechanism to be effective the nuclear quadrupole has to be in a sharp electric field gradient, a condition ideally realized by the halogen nuclei since they are univalent and therefore in a highly unsymmetrical environment, as well as at one end of a highly polar bond, usually. A nitrogen nucleus is not in quite the same type of environment since it will be attached, in many cases, to three or four ligands. A consequence of this is that the electric field gradient of the ^{14}N nucleus tends to be much less than that at a halogen nucleus, especially when there are four neighbouring atoms, so that the quadrupole interaction with the electric fluctuations, either due to the relative motion of other molecules or to molecular rotation, is small, providing therefore only an inefficient means for relaxation.

The temperature effects are explained in the following way: At very low temperatures the fluctuations are slow and will not tend to induce transitions via the quadrupole moment because there will not be a sufficiently large component of the correct frequency; therefore the lifetime of each state of the nucleus will be large and appropriate hyperfine splitting will be observed, i.e. in the proton resonance spectrum.

As the temperature rises the tumbling rate increases until it is at an optimum for inducing transitions between the quadrupole levels, i.e. there is a large component fluctuating at the nuclear quadrupole resonant frequency. The lifetime at this point will be small so that only the sharp, time-averaged effect will be observed in the proton resonance spectrum and there will be no hyperfine splitting from the nucleus which has a quadrupole.

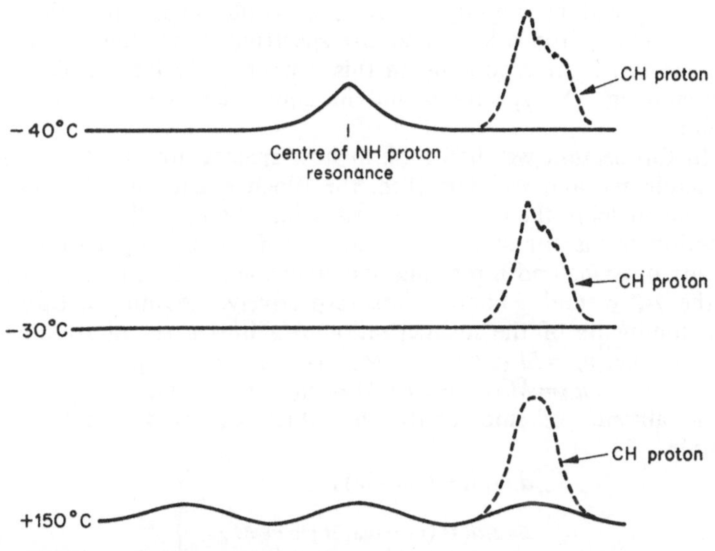

Figure 7-3. Variation of the NMR spectrum of pyrrole with temperature

At high temperatures the tumbling may be so rapid that the quadrupole experiences only the time-averaged effect of the changing electric fields; transitions are then no longer induced because the quadrupole cannot follow the rapidly varying electric field. At this point the splitting should therefore (ideally) reappear. In between the extremes the spectra will be of broad lines.

In practice all of these possibilities cannot be realized in the liquid phase, however the example of pyrrole does illustrate most of the points rather well (see Figure 7-3). Incidentally, hyperfine splitting by nitrogen and by chlorine nuclei is observed in ESR, where the time scale is that much smaller, for the splittings there are of the order of MHz rather than Hz; so evidently the lifetime of the nuclear quadrupole states is between 10^{-1} and 10^{-6} seconds.

7.3. EFFECTS OF CHANGING ENVIRONMENT OF A SPIN-INTERCHANGE BETWEEN TWO TYPES OF POSITION

If a spin could be fixed in one or other of two situations, the magnetic resonance spectrum of an assembly of such spins would consist of a simple sum of the spectra of the two types of spin. Thus if each spectrum were of a single sharp line, the spectrum of the mixture would be two sharp lines, the relative intensities depending on their relative concentration. On the other hand, if the spins were interchanging positions rapidly, each spin would experience the same average field as the others, and the spectrum would then consist of only one type, corresponding to this average. At intermediate rates of interchange the spectra would lie somewhere between these two extremes.

In this section we shall look at how spectra are affected by rates of interchange and we start from the Bloch equations. Suppose we label the nuclei in the two types of situation A and B. We shall exclude relaxation terms for simplicity, and as before the applied fields are: a steady one, B_0, and a rotating one $B' \cos \omega t$, $-B' \sin \omega t$; and these are the z-, x- and y-components respectively. Again we transform the components of the magnetization M_A into a set of rotating co-ordinates, i.e. $v_A = M_{Ay} \sin \omega t - M_{Ax} \cos \omega t$—out-of-phase

$u_A = M_{Ax} \sin \omega t + M_{Ay} \cos \omega t$—in-phase

In the absence of interchange the Bloch equations would be for example:

$$\left.\begin{aligned} du_A/dt &= (\omega_0 - \omega)\, v_A \\ dv_A/dt &= (\omega - \omega_0)\, u_A + \omega' M_{Az} \\ dM_{Az}/dt &= -\omega' v_A \end{aligned}\right\} \tag{7.7}$$

where $\omega' = \gamma B'$, $\omega_0 = \gamma B_0$, ω = angular velocity of the radiation.

These equations are not complete for they do not take into account the interchange between situations A and B. If P_{ab} is the probability per unit time of a change $A \rightarrow B$ and P_{ba} for the opposite, then the magnetization of species A, say, will change by an amount $-P_{ab}M_A$ due to some of the A spins changing into B's, and by an amount $+P_{ba}M_B$ due to B spins changing into A's. The reason why we put M_B here is that the spins which have just become A spins, will have an overall phase the same as M_B and not that of the A spins of longer standing. The reason for this difference in phase is that the A and B spins are precessing about different resultant fields, on an average.

Under steady conditions $n_A P_{ab} = n_B P_{ba}$ so that although $P_{ab}u_A \neq P_{ba}u_B$, because of the phase difference, $P_{ab}M_{Az} = P_{ba}M_{Bz}$, since $M_{Az} \propto n_A$ etc.

Eq. (7.7) becomes for the A spins:

$$\left.\begin{aligned} du_A/dt &= (\omega_A - \omega)\, v_A + P_{ba}u_B - P_{ab}u_A \\ dv_A/dt &= (\omega - \omega_A)\, u_A + \omega_A' M_A + P_{ba}v_B - P_{ab}v_A \\ dM_A/dt &= \omega_A' v_A \end{aligned}\right\} \tag{7.8}$$

where for convenience we write M_A for M_{Az}, ω_A for $(\omega_A)_0$, and so on. A similar set of equations apply for the B spins.

When steady conditions are reached, the A and B spins will lag behind the applied radiation, i.e. the rotating magnetic field, by different constant amounts. This means that the six time-derivatives, three of which are shown in Eq. (7.8), will all be zero, giving us six equations from which we obtain the following expression for $v = v_A + v_B$, the total out-of-phase component that is proportional to the absorption:

$$v = \frac{\omega'(\omega_A - \omega_B)[P_{ab}M_A(\omega - \omega_B) - P_{ba}M_B(\omega - \omega_A)]}{(\omega - \omega_A)^2\,(\omega - \omega_B)^2 + [P_{ba}(\omega - \omega_A) + P_{ab}(\omega - \omega_B)]^2} \tag{7.9}$$

A number of interesting results can be deduced from this equation. First if $P_{ab}(\omega_A - \omega_B)$ is large, the shape of the absorption curve near $\omega = \omega_A$, is given by

$$v(\omega_A) = P_{ab}M_A/((\omega - \omega_A)^2 + P_{ab}{}^2)$$

which corresponds to a line width of P_{ab}. If we had included the natural relaxation times T_{2A} and T_{2B}, the width of the line centred near $\omega = \omega_A$ would be $1/T_{2A} + P_{ab}$, so the interchange is said to give rise to what is called "lifetime broadening". In a similar way the resonance of B nuclei is broadened an amount P_{ba}.

When both P_{ab} and P_{ba} are large enough there is only one signal whose centre is given by $\omega = (\omega_A P_{ba} + \omega_B P_{ab})/(P_{ab} + P_{ba})$; i.e. situated proportionately in between the two signals of the isolated A and B species.

When these probabilities of interchange are not too large, however, there will be two lines, and since $P_{ab} \propto 1/n_A$, the line corresponding to the more numerous species will be not only sharper, but also much taller (height $\propto n_A{}^2$). It is difficult to find an example where there is simple interchange between two situations which have different concentrations, however there is one case which has been observed, and that is an exchange of protons between tert-butyl alcohol and water in acetone solution, where as long as the water concentration is kept low, a differential line broadening effect can be seen (see Figure 7-4).

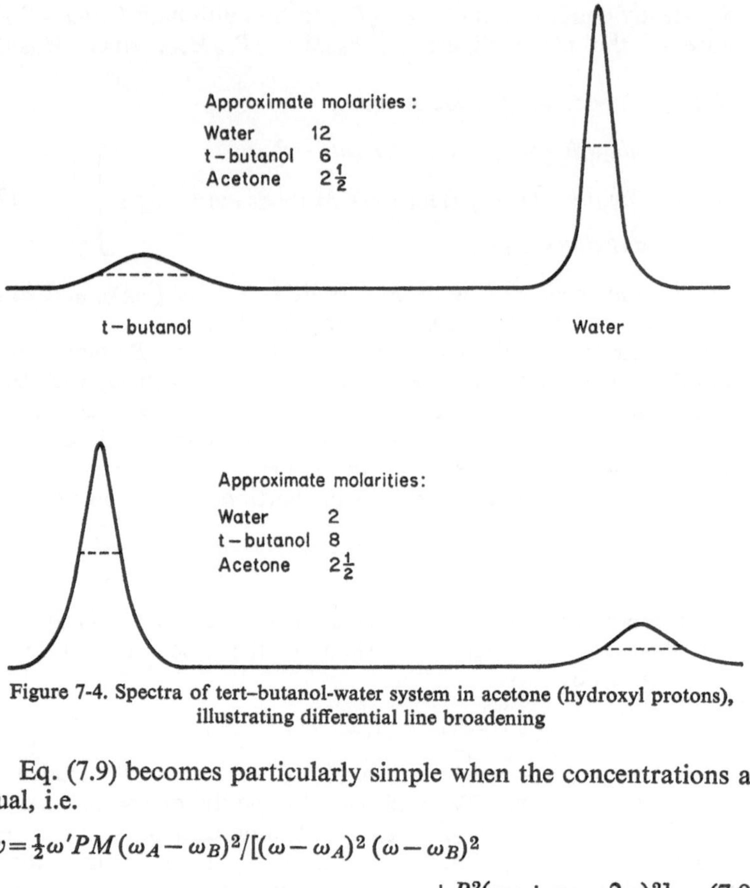

Figure 7-4. Spectra of tert–butanol-water system in acetone (hydroxyl protons), illustrating differential line broadening

Eq. (7.9) becomes particularly simple when the concentrations are equal, i.e.

$$v=\tfrac{1}{2}\omega' PM(\omega_A-\omega_B)^2/[(\omega-\omega_A)^2(\omega-\omega_B)^2 + P^2(\omega_A+\omega_B-2\omega)^2] \quad (7.9a)$$

$$P=P_{ab}=P_{ba};\quad M_A=M_B=M/2$$

The extreme values of this function occur at

$$=\tfrac{1}{2}(\omega_A+\omega_B)\pm\sqrt{(\omega_A-\omega_B)^2-8P^2}$$

which are maxima, subject to the condition $(\omega_A-\omega_B)^2>8P^2$ and at

$$\tfrac{1}{2}(\omega_A+\omega_B)$$

—which is a minimum if $(\omega_A-\omega_B)^2>8P^2$ and a maximum otherwise.

A good example where the interchange can be speeded up by simply raising the temperature is in the isomerization of N,N-dimethylnitrosamine (see Figure 7-5).

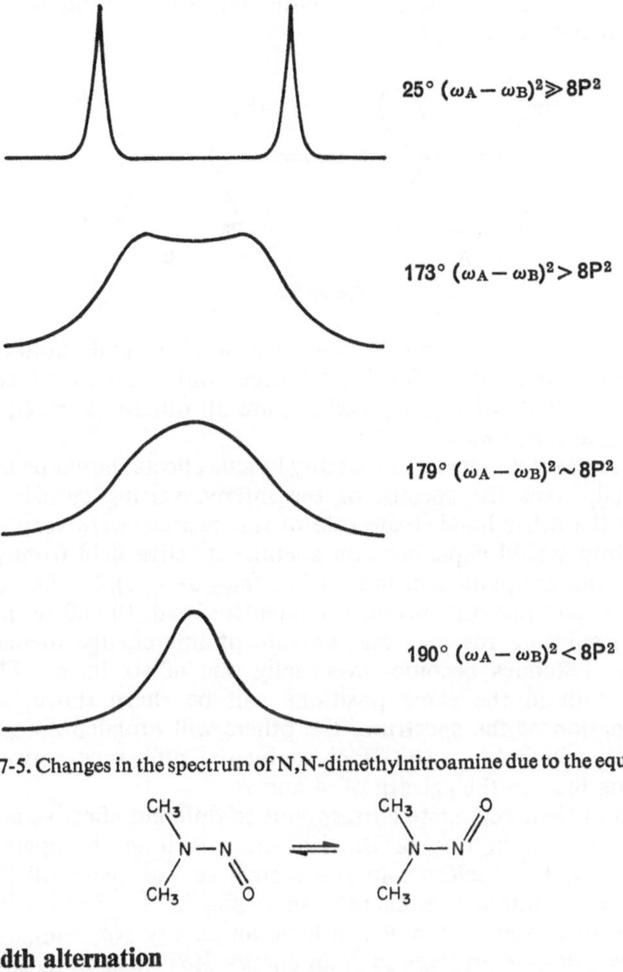

Figure 7-5. Changes in the spectrum of N,N-dimethylnitroamine due to the equilibrium.

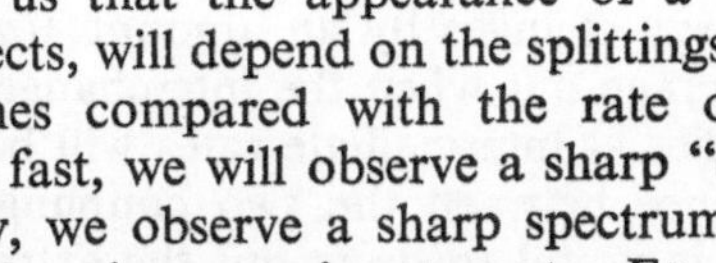

Line-width alternation

Eq. (7.9) tells us that the appearance of a spectrum in which there are kinetic effects, will depend on the splittings $(\omega_A-\omega_B)$ between the appropriate lines compared with the rate of the interchange. When the rates are fast, we will observe a sharp "average" spectrum; when they are slow, we observe a sharp spectrum which is a simple sum of those of the various species present. Eq. (7.9) indicates how to deal with intermediate situations, and some examples of these occur in the ESR spectra of certain radicals, giving rise to a phenomenon called line-width alternation. We shall look at two cases of slightly different orders of complexity.

Example: The vinyl radical apparently undergoes an internal change of conformation of the type:

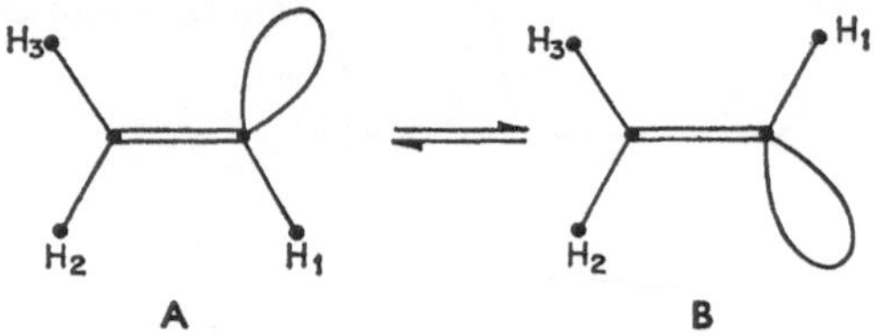

Figure 7-6.

The ESR spectra of *A* and *B* would be identical, and should consist of 8 lines of equal intensities due to three doublet splittings, since the coupling constants of H_1, H_2 and H_3 are all different; in A_1, $a_1 = a_\alpha$, $a_2 = a_{,\text{trans}}$ and $a_3 = a_{\text{cis}}$.

At first sight it seems odd that any kinetic effects should be observed here at all, since the spectra of the interconverting species are the same, on the other hand if the rate of this process were very fast, the odd electron would experience an average effective field from protons 2 and 3, the coupling constant being $(a_{\text{trans}} + a_{\text{cis}})/2$. Under these conditions the protons become equivalent and therefore a triplet (1 : 2 : 1) splitting results. As the rate of interchange increases the spectrum of 8 lines becomes eventually one of six lines. The lines which remain in the same positions will be sharp throughout the transformation of the spectrum, the others will broaden according to Eq. (7.9a), which $(\omega_A - \omega_B)$ is given by the difference between corresponding lines in the spectra of *A* and *B*.

Each nuclear spin state corresponds to different effective acting on the odd electron in one of the isomers. Writing the spins in the order 1, 2, 3; the nuclear spin states such as $\alpha\alpha\alpha$, $\beta\alpha\alpha$, will have the same hyperfine interaction energies in *A* and in *B*. The nuclear spin states $\alpha\beta\alpha$ in *A* and $\alpha\alpha\beta$ in *B* will have an energy $\frac{1}{4}(a_\alpha - a_{\text{trans}} + a_{\text{cis}})$, and states $\alpha\alpha\beta$ in *A* and $\alpha\beta\alpha$ in *B* an energy $\frac{1}{4}(a_\alpha - a_{\text{cis}} + a_{\text{trans}})$ so that when the nuclear spin state of *A* is $\alpha\beta\alpha$, and it flips over to *B* the hyperfine interaction energy changes by an amount $\frac{1}{2}(a_{\text{trans}} - a_{\text{cis}})$. These are states which average out when the interchange is fast enough and the shapes of the lines at intermediate rates will be determined by the ratio of the difference between the two coupling constants and the reciprocal rate. It is useful to draw out the extreme "stick" spectra, labelling the various lines according to the corresponding nuclear spin states, since then we can easily see which lines may become broad.

After allowing for the doublet splitting due to H_1, we can see that this is an example of line-width alternation, i.e. the two outer

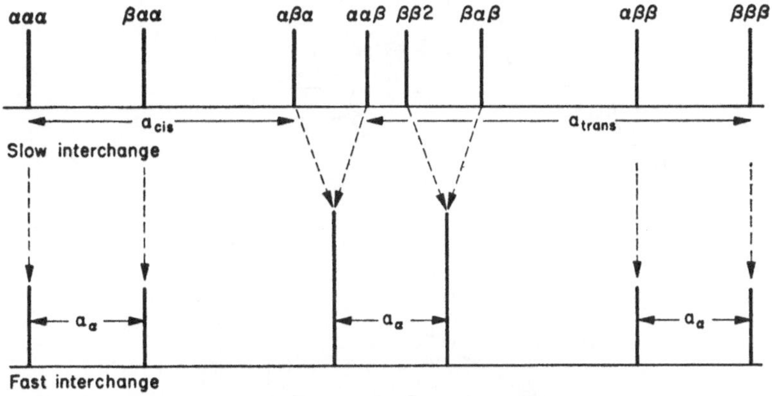

Figure 7-7. Kinetic effects in ESR spectrum of vinyl.

pairs of lines are sharp but the inner pairs become broad at intermediate rates; in fact when the differences in coupling constants are large, the centre lines disappear altogether from the spectrum and one sees only the outer pairs.

Example: The Cyclohexyl radical: Generally speaking, six-membered rings such as cyclohexane derivatives, dioxane, etc., exist in which is called a chair conformation, because in this arrangement steric strain is at a minimum. In the absence of bulky substitutes there is usually a low potential barrier between the two possible chair conformers, and therefore the rings are continually flapping over from one to the other. The effect of this flapping is to interchange axial and equatorial protons, for a proton which is axial in one conformation of, say, a cyclohexyl radical, will become equatorial when the ring flips over to the alternative conformation (see Figure 7-8).

Leaving out the effects of the α proton for the moment, since all it does is to double the number of lines in the ESR spectrum, there are two axial and two equatorial protons. If there were no interchange of conformations, this would result in two triplet (1 : 2 : 1) splittings. When the interchange is rapid enough, all four protons become effectively equivalent, on an average, and the spectrum would yield correspondingly, a quintet (1 : 4 : 6 : 4 : 1).

At intermediate flapping frequencies a glance at Figure 7-9 shows us which lines should become broad. The only lines which are sharp over the whole range are the two outermost ones. The centre line too will be sharp over most of the range of flapping frequencies, because of the wide difference in resonant frequency of the other two lines which have eventually to coalesce to give the centre line of the fast flapping spectrum an intensity of six units. At intermediate

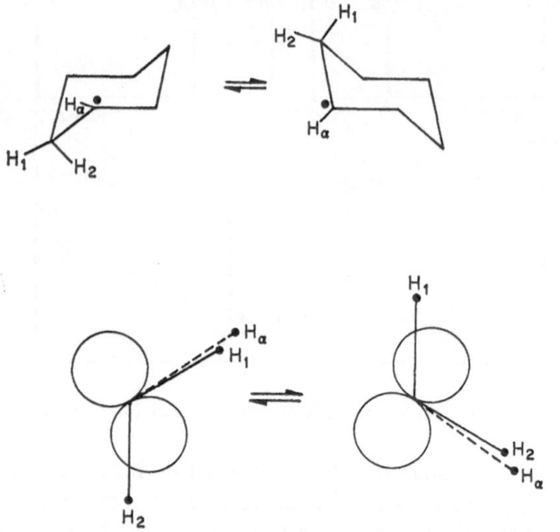

Figure 7-8. Ring flapping in cyclohexyl and related radicals.

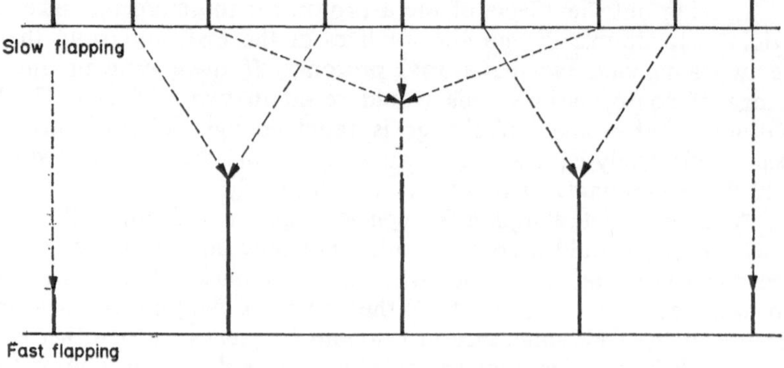

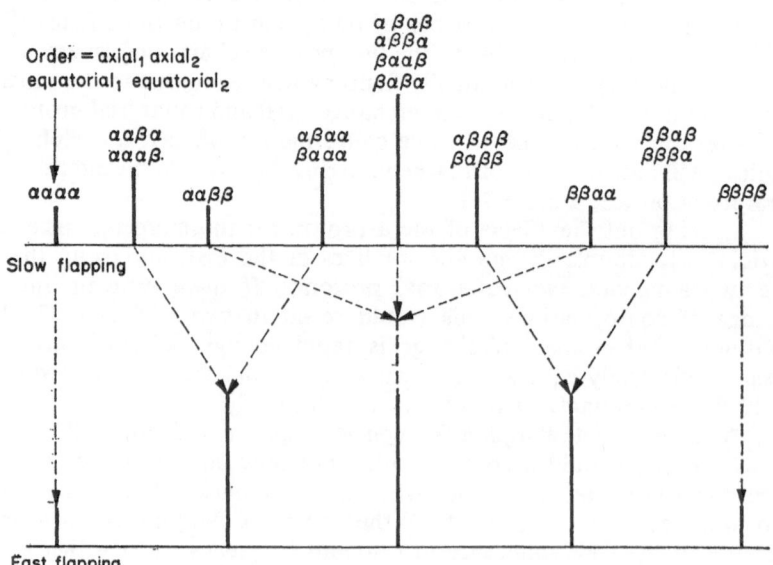

Figure 7-9. Theoretical spectra of cyclohexyl neglecting α-splitting.

rates of flapping, then, alternate lines are either sharp or broad, (e.g. see Figure 7-10).

From the example given we can see that the splitting from the β-protons is very nearly 1 : 4 : 1, because the other absorptions are so broad.

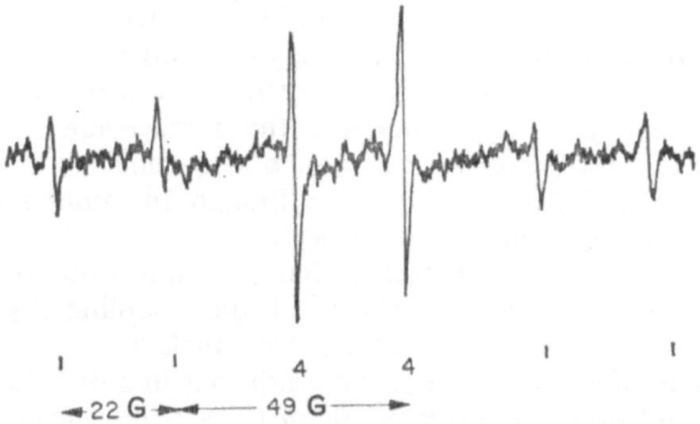

Figure 7-10. ESR spectrum of $\dot{H}C\langle{CH_2CH_2 \atop CH_2CH_2}\rangle NH$.

7.4. DOUBLE RESONANCE

We have now established the effects of the "average" environment on the resonance of a spin, i.e. the interactions of the spin must give a well-defined average value over a certain time scale and they must be either varying very rapidly or very slowly for there to be a sharp resonance. In the case of ethanol, the resonance of the methylene protons is split into two by the hydroxyl proton, and that of the hydroxyl proton is a triplet due to the interaction with the two methylene protons. When acid is added a rapid exchange is set up amongst the hydroxyl and the acid protons. From the point of view of the methylene protons, the nearby hydroxyl proton keeps changing, and with it the spin, i.e. at one instant there might be an α-spin interacting with the methylene protons, and at the next instant it is replaced by a proton with β-spin. When the exchange is fast enough the overall average coupling with the hydroxyl proton will be zero—so the corresponding splitting of the methylene proton resonance will disappear. From the point of view of the hydroxyl proton, the neighbouring methylene group, and the corresponding spin states of its protons, changes with the exchange process, and therefore the splitting of the hydroxyl proton resonance will disappear too.

There are of course other ways by means of which the spin of a neighbouring nucleus can change, or at least effectively change (one example being by nuclear quadrupole relaxation), and provided the chemical shifts are sufficiently different, one convenient way of getting this to happen deliberately, is to irradiate the sample strongly at the resonance frequency of the second nucleus. There are several ways of looking at the results of doing this, one being that the resonance of nucleus 2, say, is saturated, i.e. very rapid transitions between its levels are taking place, and therefore the average effect at the first nucleus will be zero and the splitting due to nucleus 2 disappears. This is called spin decoupling, and although the final result is the same, we cannot use the results of section 7.3 in dealing with intermediate cases, i.e. when the rate of induced transitions of nucleus 2 is not sufficient to make the effects of the coupling J_{12} disappear altogether from the resonance spectrum of nucleus 1.

The main difference between the conditions in a double irradiation experiment and one where transitions of the second nucleus are being induced by some random process, is that both nuclei are "bathed" in the same radiation, which means that the first nucleus is in a resultant field consisting of the steady applied field plus that of the radiation, which for simplicity we can regard as rotating at the resonance frequency of the second nucleus.

The major difference between this form of "induced" relaxation and the others we have discussed is that the lines in the intermediate stages remain sharp; only the extreme cases look the same as in other forms of relaxation.

Double resonance is used in NMR in two main ways, one being when the secondary radiation is applied with relatively low intensity at the frequency of one of the lines in the X-spectrum whilst the A-spectrum is being observed, i.e. in an AX compound. In this case only the lines which are derived from the transitions involving either of the two states between which the X-transition occurs, are affected. This is known, amusingly, as a tickling experiment.

The other way of employing double resonance, usually simply for the purpose of simplifying the A spectrum, is to apply intense irradiation at the centre frequency of the X-spectrum, when as we have already discussed, all coupling from X disappears in the A spectrum.

In the spin tickling experiment the affected peaks get split into two and by noting which peaks change on applying the radiation of frequency of say the lowest field peak of nucleus 1, relative signs of various coupling constant can be deduced. A simple example best illustrates the point: in the spectrum of three nuclei $X_1X_2X_3$, the

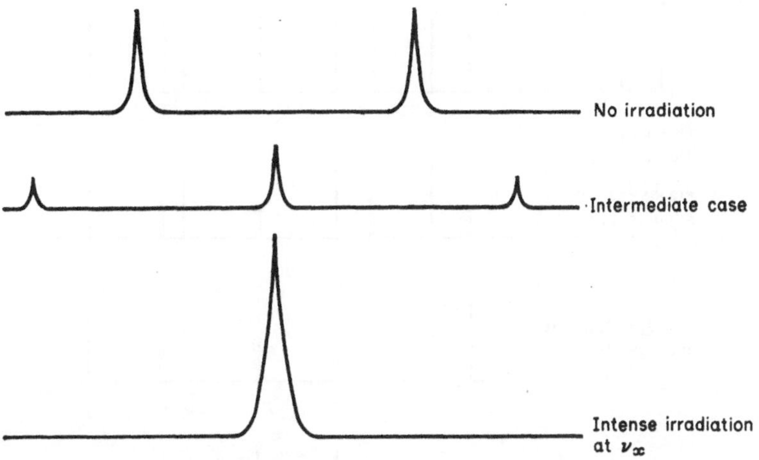

Figure 7-11. Effect of irradiation at X frequency on A part of AX spectrum.

transitions for nucleus 1:

$$\begin{pmatrix} \alpha\alpha\alpha \leftrightarrow \beta\alpha\alpha \\ \beta\beta\beta \leftrightarrow \alpha\beta\beta \\ \alpha\beta\alpha \leftrightarrow \beta\beta\alpha \\ \alpha\alpha\beta \leftrightarrow \beta\alpha\beta \end{pmatrix} \begin{matrix} \nu_1 - \frac{1}{2}(J_{12} + J_{13}) \longleftarrow \text{“tickle” this line} \\ \nu_1 + \frac{1}{2}(J_{12} + J_{13}) \\ \nu_1 + \frac{1}{2}(J_{12} - J_{13}) \\ \nu_1 - \frac{1}{2}(J_{12} - J_{13}) \end{matrix}$$

for nucleus 2:

$$\begin{pmatrix} \alpha\alpha\alpha \leftrightarrow \alpha\beta\alpha \\ \beta\beta\beta \leftrightarrow \beta\alpha\beta \\ \beta\alpha\alpha \leftrightarrow \beta\beta\alpha \\ \alpha\alpha\beta \leftrightarrow \alpha\beta\beta \end{pmatrix} \begin{matrix} \nu_2 - \frac{1}{2}(J_{12} + J_{23}) \longleftarrow \\ \nu_2 + \frac{1}{2}(J_{12} + J_{23}) \\ \nu_2 + \frac{1}{2}(J_{12} - J_{23}) \longleftarrow \\ \nu_2 - \frac{1}{2}(J_{12} - J_{23}) \end{matrix}$$

for nucleus 3:

$$\begin{pmatrix} \alpha\alpha\alpha \leftrightarrow \alpha\alpha\beta \\ \beta\beta\beta \leftrightarrow \beta\beta\alpha \\ \beta\alpha\alpha \leftrightarrow \beta\alpha\beta \\ \alpha\alpha\beta \leftrightarrow \alpha\beta\alpha \end{pmatrix} \begin{matrix} 3 - \frac{1}{2}(J_{13} + J_{23}) \longleftarrow \\ 3 + \frac{1}{2}(J_{13} + J_{23}) \\ 3 + \frac{1}{2}(J_{13} - J_{23}) \longleftarrow \\ 3 - \frac{1}{2}(J_{13} - J_{23}) \end{matrix}$$

these lines are affected

Alternate lines in the spectra of 2 and 3 will be affected; if in both cases it is the line at lowest field which is affected, all of the coupling constants J_{12}, J_{13} and J_{23} must have the same sign. If the lowest line of spectrum 2 is not affected, then J_{12} and J_{13} will have opposite signs, and so on (see Figure 7-12).

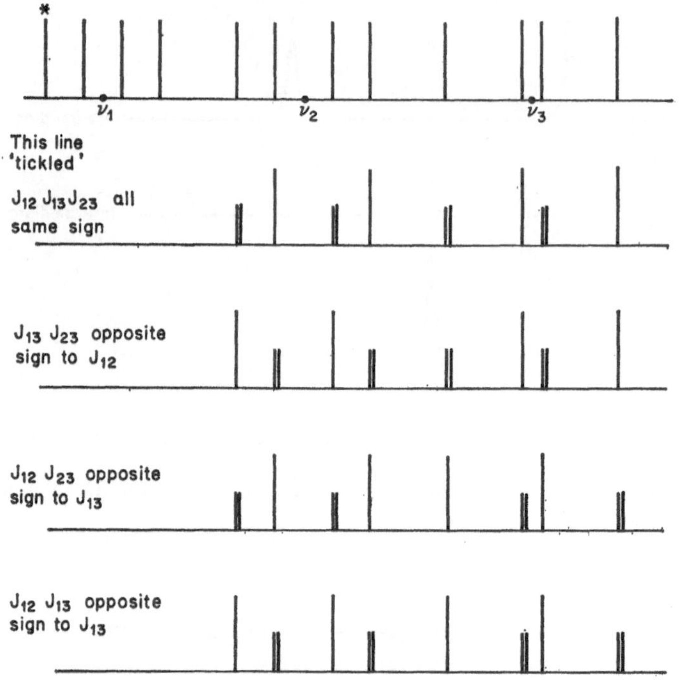

Figure 7-12. Possible results of irradiating lowest field line of X_1 in spectra of X_2X_3.

7.5. ELECTRON–NUCLEAR DOUBLE RESONANCE

The coupling of a nucleus with odd electrons generally produces what we can loosely call a Knight shift in its NMR spectrum, even when the relaxation of the electron spins is very fast, because of the unequal populations of the electron spin magnetic energy levels. Depending on the relaxation conditions various types of experiment have been designed around the hyperfine interactions between odd electrons and nuclei, involving irradiation of the sample at both NMR and ESR frequencies. We shall look at two types which for a given sample are mutually exclusive.

The ENDOR Experiment

When the nuclei relax independently of the odd electrons although they are coupled to them (no Overhauser effect), it should be possible to conduct an ENDOR (electron–nuclear double resonance) experiment as follows:

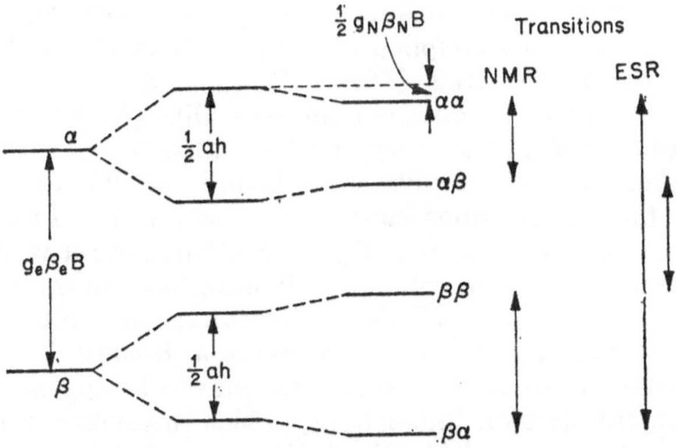

Figure 7-13. Energy level scheme for one odd electron–one nucleus, spin (2)½, coupling constant *a*.

The first step is to obtain the ESR spectrum, which may be rather poorly resolved. In our simple example (see Figure 7-13) one of the electron resonance lines, or part of the spectrum, is selected and the field and frequency kept at their corresponding values. The power is then increased so that the point of saturation is reached and the intensity of the absorption decreases to zero, i.e. the signal disappears. What we have done is to equalize the populations of the two levels involved in the transition, in our case say $\beta\alpha$ and $\alpha\alpha$.

Now a second r.f. field is applied to the sample and its frequency swept between, say, 0→100 MHz. If this secondary radiation is intense enough then at a frequency given by $h\nu = \frac{1}{2}ah - g_N\beta_N B$, the transitions between $\alpha\beta$ and $\alpha\alpha$ will become saturated, and consequently the balance between states $\alpha\alpha$ and $\beta\alpha$ will be disturbed and the ESR signal will reappear and then decrease again to zero as saturation of the electron resonance sets in once more. A relatively accurate value of the coupling constant can be obtained by taking the mean of the two frequencies, $\frac{1}{2}a \pm g_N\beta_N B/h$, which give a visible effect.

Where applicable, this technique is a good one for measuring coupling constants because lines which appear close together on a scale of, say, 10,000 MHz in the ESR spectrum, correspond to lines which are well-separated on a scale of 100 MHz.

The Overhauser Effect

When the coupling between the odd electron and a nucleus is time-dependent, their relaxation processes are connected and there is the possibility of observing an Overhauser effect. In this discussion we are interested in introducing mechanisms for double spin flips! that is, in transitions in which both electron and nuclear spins change direction. There are two distinct ways by which this can take place, first, there is the scalar coupling $aS.I$ which connects the levels $\alpha_e\beta_N$ and $\beta_e\alpha_N$, due to the terms $\frac{1}{2}(S_+I_- + S_-I_+)$; and secondly there are the dipolar interactions, which contain terms like $(\bar{S}.\bar{r})(\bar{I}.\bar{r})$ and which can therefore connect states $\alpha_e\alpha_N$ and $\beta_e\beta_N$ as well.

For them to be effective in inducing the "double-quantum" transitions these interactions have to vary with time. In the case of the dipolar interaction this is easily achieved because it is anisotropic so that simple molecular rotation will introduce the required time-dependence. In the case of the scalar interaction, rotation of the molecule will not give rise to any fluctuation, because it is isotropic. If the scalar terms are to vary, then, there has to be appropriate vibrations or intermolecular interactions, which introduce time-varying spin densities on the nuclei. We shall assume that these interactions are frequent enough for them to be significant in determining the steady state conditions in the following discussion.

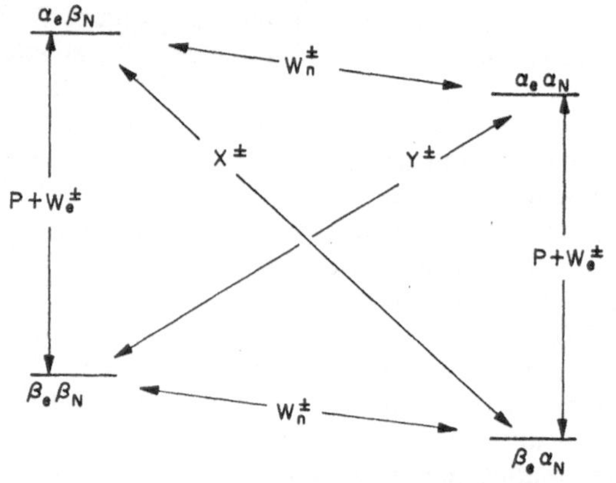

Figure 7-14. Transitions involved in Overhauser effect

In Figure 7-14 the convention is that a label + is for an upward transition and − for a downward one, labels n, e are for the nucleus and electron respectively, and for convenience we shall drop these labels in the wavefunctions and always write the electronic one first. $W_n^{\pm}$, $W_e^{\pm}$, $X^{\pm}$, $Y^{\pm}$ are appropriate transition probabilities due to relaxation processes and P the transition probability due to the radiation applied at the ESR frequency.

We assume that there is no average coupling over a period of time, so that the hyperfine interactions which give rise to the relaxation processes represented by $X^{\pm}$ and $Y^{\pm}$ fluctuate about zero; this is why the order of the energy levels is that given.

Having thus defined the conditions of the experiment we say what the Overhauser effect is, i.e. when the electron resonance is saturated (P very large), the populations of the nuclear spin energy levels may be affected in such a way that the intensity of the NMR absorption is greatly increased.

The problem is then to find the relative populations of the nuclear spin states, which will be defined by the ratio $(n_{\alpha\alpha}+n_{\beta\alpha})/(n_{\beta\beta}+n_{\alpha\beta})$. Under steady conditions, which are only equilibrium conditions when P is zero, the populations remain constant with time, hence for level $\alpha\beta$

$$n_{\alpha\beta}(P+W_e^-+X^-+W_n^-)=n_{\alpha\alpha}W_n^++n_{\beta\alpha}X^++(P+W_e^+)\,n_{\beta\beta} \quad (7.10)$$

and for level $\beta\beta$

$$n_{\beta\beta}(P+W_e^++Y^++W_n^-)=n_{\beta\alpha}W_n^++n_{\alpha\alpha}Y^-+(P+W_e^-)\,n_{\alpha\beta}$$

adding these two equations we remove P and $W_e^\pm$:

$$n_{\beta\alpha}(X^++W_n^+)+n_{\alpha\alpha}(W_n^++Y^-)=n_{\alpha\beta}(W_n^-+X^-)$$
$$+n_{\beta\beta}(W_n^-+Y^+) \quad (7.11)$$

When the electron resonance is saturated, levels connected by a transition become equally populated, i.e. $n_{\alpha\alpha}=n_{\beta\alpha}$, and $n_{\beta\beta}=n_{\beta\alpha}$; and the ratio for the nuclear levels becomes:

$$(n_{\alpha\alpha}+n_{\beta\alpha})/(n_{\beta\beta}+n_{\alpha\beta})=(2W_n^-+X^-+Y^+)/(2W_n^++X^++Y^-) \quad (7.12)$$

The various transition probabilities can be estimated relative to each other from the populations in the absence of radiation, for the numbers in the levels are then determined by Boltzmann's law. It is sufficient for our purpose to look at some extreme cases.

(i) $X^\pm$, $Y^\pm$ are all small relative to $W_n^\pm$.

In this case the population ratio $n_{\alpha\alpha}/n_{\beta\beta}=W_n^-/W_n^+$, and is therefore the same as in a single irradiation experiment.

(ii) If $X^\pm$ is much larger than the others the ratio when there is saturation is $n_{\beta\alpha}/n_{\alpha\beta}=X^-/X^+$ $(=n_{\alpha\alpha}/n_{\beta\beta})$.

But in this case we obtain from Eq. (7.11) $n_{\beta\alpha}/n_{\alpha\beta}=X^-/X^+$, i.e. this applies even in the absence of the electron resonance radiation, e.g. under equilibrium conditions when

$$n_{\beta\alpha}{}^0/n_{\alpha\beta}{}^0=\exp\,(g_n\beta_n+g_e\beta_e)B/kT.$$

Since this ratio is the same as the ratio of the populations of the two nuclear magnetic energy levels when saturation conditions prevail, and since normally in NMR this ratio is $\exp\,(g_n\beta_nB/kT)$ we can see there has been a considerable change in the population difference of the nuclear states. This increase in population difference in the nuclear levels will result in a corresponding increase in the intensity of absorption, the ideal enhancement factor being

$$(g_n\beta_n + g_e\beta_e)/g_n\beta_n \sim 660.$$

$X^{\pm}$ is only large when there is a time-varying contact (scalar) interaction, and one example where this happens is in solutions of alkali metals in liquid ammonia, where the odd electron moves rapidly from one molecule to another, thus producing a suitable fluctuation in the hyperfine interactions with the various protons and nitrogen nuclei. The proton resonance absorption here increases several hundredfold when the electron resonance is saturated.

(iii) If $Y^{\pm}$ is larger than the other transition probabilities in Eqs. (7.11) and (7.12) we can write

$$n_{\beta\beta}/n_{\alpha\alpha} = Y^-/Y^+ = n_{\beta\beta}{}^0/n_{\alpha\alpha}{}^0 = \exp\,(g_e\beta_e - g_n\beta_n)\,B/kT.$$

Now we have exactly the reverse of the previous situation, for the state for nuclear spins has a population determined by the ratio $n_{\alpha\alpha}/n_{\beta\beta}$ which is less than unity. The populations are therefore reversed on saturation for there will be more in the state of β nuclear spin than in the two levels with α nuclear spin.

Under the usual conditions of an NMR experiment

$$\frac{\text{number of } \beta \text{ nuclear spins}}{\text{number of } \alpha \text{ nuclear spins}} = \exp\,(-g_n\beta_n B/kT)$$

So there is a negative enhancement factor (i.e. emission instead of absorption) whose ideal value is approximately

$$\left|1 - \frac{g_e\beta_e}{g_n\beta_n}\right|.$$

In practice $Y^{\pm}$, which is introduced by dipolar relaxation, is always accompanied by relaxation mechanisms of the $X^{\pm}$ and $W_n{}^{\pm}$ types so the negative enhancement factors are not as large as those discussed in (ii) above. However, in concentrated solutions of the naphthalene negative ion enhancements of up to -60 have been observed.

Finally it is worthwhile to mention that time-varying coupling between two nuclear spins can also give the right relaxation conditions so that saturation of one of their resonances can lead to an enhancement of the absorption of the other.

Chemically Induced Dynamic Nuclear Polarization (CIDNP)

From one point of view, the relaxation times of nuclei reflect their "memory" of what has happened to them. If, for some reason, their spin populations are disturbed, some time will elapse before they have "forgotten" it, i.e. before they have regained their steady-state populations.

The Overhauser effect, which we have just been discussing, depends on the non-equilibrium condition of the electron spins, i.e. the two electronic spin states are equally populated instead of having their populations determined by Boltzmann's law. In fact if we could arrange conditions so that the electronic spin states are equally populated, it might then be possible to observe an Overhauser effect, provided that there is the right sort of coupling between electronic and nuclear spins. In other words we might be able to observe an enhancement of the NMR absorption.

One such state of affairs exists when a molecule dissociates into two radical fragments

$$R-R \rightarrow 2R\cdot = R\uparrow + R\downarrow$$

Immediately after dissociation the odd electron spin states will be equally populated and we may observe an Overhauser effect in the NMR spectrum of the free radical $R\cdot$ until the electron spins have reached their equilibrium populations.

Another way is to "select" equal numbers of opposite spins by more or less the reverse process.

$$R\uparrow + R\downarrow \rightarrow R-R$$

In the group of radicals destined to dimerize, the nuclear spin populations may be polarized. This would not be apparent in the NMR spectrum of the free radicals because there would be a residual polarization of the other radicals, which would counterbalance it.

However, in the product, the nuclear spin populations will remain polarized until relaxation processes have acted to give the usual steady state distribution. An enhanced spectrum may therefore be observed, i.e. of the product, which will decay to the ordinary spectrum after a time determined by the relaxation conditions.

$R\cdot$ radicals whose spin populations are given by Boltzmann's law

↙ ↘

$R\cdot$ equal populations (Overhauser effect) — $R\cdot$ remainder (not reacting) (electron spins polarized)

↓

R_2 product, nuclear populations still polarized, hence enhanced spectrum

↓

R_2 after several relaxation times nuclear populations given by Boltzmann law—normal spectrum.

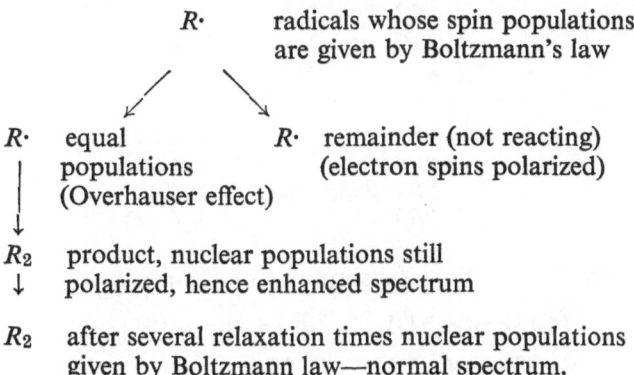

Example: Photolysis of benzophenone in alkyl benzenes ϕ CH_2R

$$\phi_2\,C{=}O \ldots \rightarrow \phi_2\,COH + \phi\,CH^*R \text{ (triplet state)}$$

dynamic polarization step

$$\downarrow$$

$$\phi_2\,C(OH)\,CH^*R\phi$$

enhanced NMR spectrum of H*

Nuclear spin levels, $R \equiv H^*$

$\beta\beta m_z = -1$, population n_{-1}

$(\alpha\beta \pm \beta\alpha)/\sqrt{2} m_z = 0$, population n_0 absorption $\propto n_0 + n_{+1}$

$\alpha\alpha m_z = +1$, population n_{+1} emission $\propto n_0 + n_{-1}$

hence intensity of absorption $\propto (n_{+1} - n_{-1})$

observed enhanced spectrum due to "energy polarization".

In contrast to this the transitory enhanced spectrum observed when the substrate is ethyl benzene is apparently largely due to entropy polarization. Product ϕ_2 C(OH) $CH^*(CH_3)\phi$. By entropy polarization we mean that there is no change in the total Zeeman energy of the proton but just compensating changes in levels $m_z = \pm m$, compared with conditions in the molecules which have existed for a relatively long time.

Nuclear spin levels

(H*)	(CH_3)	Population
β	$\beta\beta\beta$	$n_\beta(-2)$
	$\alpha\beta\beta$, $\beta\beta\alpha$, $\beta\alpha\beta$	$n_\beta(-1)$
	$\alpha\alpha\beta$, $\alpha\beta\alpha$, $\beta\alpha\alpha$	$n_\beta(0)$
	$\alpha\alpha\alpha$	$n_\beta(+1)$
α	$\alpha\alpha\alpha$	$n_\alpha(+2)$
	$\alpha\beta\alpha$, $\alpha\alpha\beta$, $\beta\alpha\alpha$	$n_\alpha(+1)$
	$\beta\beta\alpha$, $\beta\alpha\beta$, $\alpha\beta\beta$	$n_\alpha(0)$
	$\beta\beta\beta$	$n_\alpha(-1)$

Intensities of transitions (absorption) of H* are

$$(n_\alpha(2) - n_\beta(+1)), \quad 3(n_\alpha(+1) - n_\beta(0)),$$
$$3(n_\alpha(0) - n_\beta(-1)), \quad (n_\alpha(-1) - n_\beta(-2)).$$

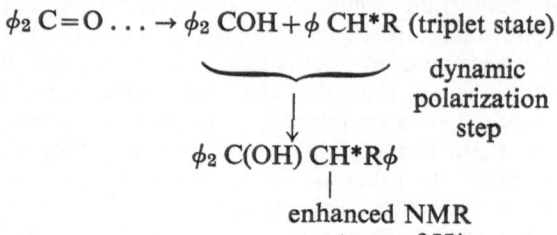

Suppose now that

$$\left.\begin{array}{c} n_\alpha(+2)=N_\beta(-2)=n_2 \\ n_\alpha(\pm 1)=N_\beta(\pm 1)=n_1 \\ n_\alpha(0)=n_\beta(0)=n_0 \end{array}\right\} \text{(i.e. for the enhanced part of the spectrum)}$$

Then the intensities of absorption will be proportional to

$$(n_2-n_1), \qquad 3(n_1-n_0), \qquad 3(n_0-n_1), \qquad (n_1-n_2).$$

i.e. if $n_0>n_1>n_2$ enhanced spectrum will look like that in Figure 7-15 which approximates to the transitory observed spectrum.

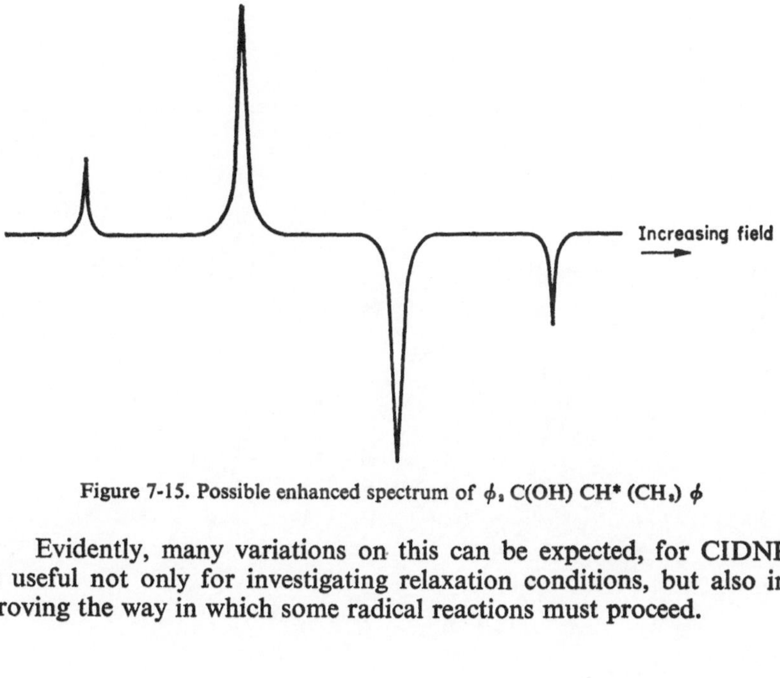

Figure 7-15. Possible enhanced spectrum of ϕ_2 C(OH) CH* (CH$_2$) ϕ

Evidently, many variations on this can be expected, for CIDNP is useful not only for investigating relaxation conditions, but also in proving the way in which some radical reactions must proceed.

Appendix

Vector Algebra

Scalars

Anything which can be characterized by a single number is called a scalar quantity for example,

The population of London
Price of a return ticket to Bournemouth
Energy
Entropy

Vectors and Tensors

Something which needs more than one number in order to be characterized may be a vector, or more generally, a tensor quantity.

Examples:

Position, velocity Force, torque Momentum, angular momentum	Vectors (tensors of rank 1)
Connection between stresses and strains Electromagnetic force Moment of inertia	Tensors of rank 2

The thing which distinguishes vectors and tensors from any arbitrary collection of numbers is the way they change their form when expressed in terms of different coordinate systems, i.e. their transformational properties. A detailed study of this would lead us into tensor calculus. However, it is sufficient for our purposes to think in geometrical terms, so that in three dimensions we say that a vector can be represented by three coordinates which express its length and direction.

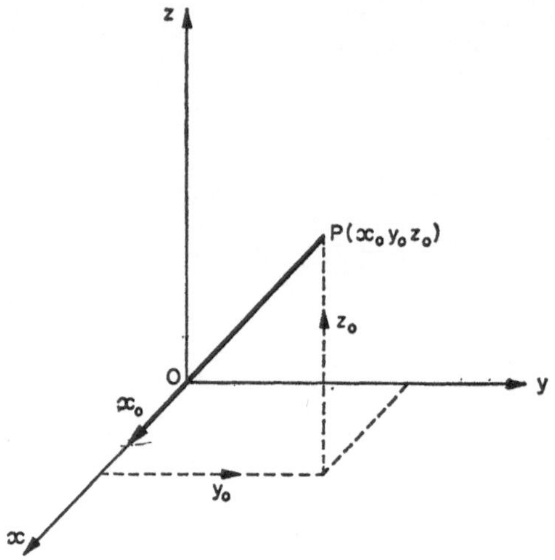

Figure A-1.

The most important property of vectors from our point of view is that they can be added together to give a resultant vector of the same type. In particular each vector can be represented by a sum of three mutually perpendicular vectors.

Thus is $\bar{i}$, $\bar{j}$, $\bar{k}$ are three unit vectors parallel to x, y and z axes respectively

$$\overline{OP} = x\bar{i} + y\bar{j} + z\bar{k}$$

If we wish to add two vectors (x, y, z) and (a, b, c) we can simply express them in terms of our three unit vectors, then

$$\begin{aligned}(x, y, z) + (a, b, c) &= x\bar{i} + y\bar{j} + z\bar{k} + a\bar{i} + b\bar{j} + c\bar{k} \\ &= (x+a)\bar{i} + (y+b)\bar{j} + (c+z)\bar{k} \\ &= (x+a,\ y+b,\ z+c)\end{aligned}$$

This is very useful for it means that we can resolve vectors into their components along convenient axes.

Scalar Product of two vectors $\bar{A}.\bar{B}$ is defined as the product of their magnitudes $\times$ cosine of the angle between them, thus

$$\begin{aligned}\bar{A}.\bar{B} &= (a_x\bar{i} + a_y\bar{j} + a_z\bar{k}).(b_x\bar{i} + b_y\bar{j} + b_z\bar{k}) \\ &= a_xb_x\bar{i}.\bar{i} + a_yb_y\bar{j}.\bar{j} + a_zb_z\bar{k}.\bar{k}\end{aligned}$$

Since other terms are zero because $\bar{i}\,\bar{j}\,\bar{k}$ are mutually perpendicular to each other ($\cos \pi/2=0$)

$$\bar{A}.\bar{B}=a_xb_x+a_yb_y+a_zb_z$$

We can check this result without introducing the unit vectors $\bar{i}$, $\bar{j}$, $\bar{k}$,

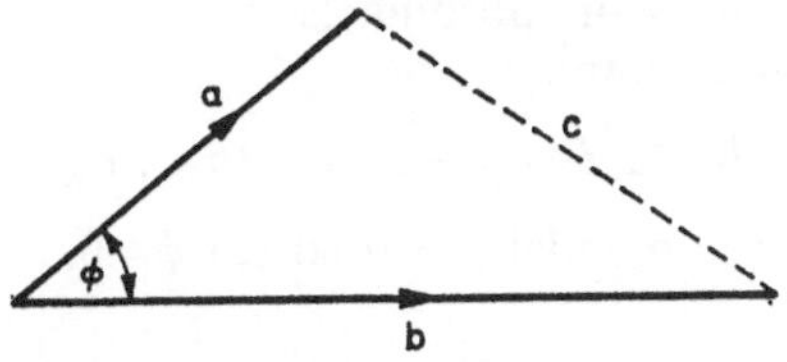

Figure A-2.

for

$$a^2+b^2-2ab\cos\phi=c^2$$

$$a^2=a_x{}^2+a_y{}^2+a_z{}^2$$

$$b^2=b_x{}^2+b_y{}^2+b_z{}^2$$

$$c^2=(a_x-b_x)^2+(a_y-b_y)^2+(a_z-b_z)^2$$

Hence

$$2ab\cos\phi=2a_xb_x+2a_yb_y+2a_zb_z$$

i.e.

$$ab\cos\phi=\bar{A}.\bar{B}=a_xb_x+a_yb_y+a_zb_z$$

Vector Product

The vector product of two vectors is another vector which is perpendicular to both of them and whose magnitude is given by the product of their lengths multiplied by the sine of the angle between

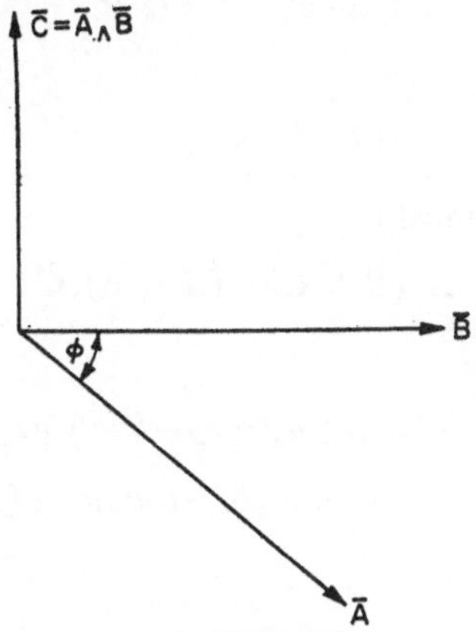

Figure A-3.

them. Since the result of a vector product is a vector we have to be careful about our definition of its direction.

We define the product $\bar{A} \wedge \bar{B}$ so that its direction is such that looking along it, the original vector $\bar{B}$ is in a clockwise sense with respect to $\bar{A}$ (see Figure A-3).

The most useful form of the "vector" or cross-product is that given in terms of cartesian coordinates.

Using our unit vectors:

$$\bar{A} \wedge \bar{B}=(a_x i+a_y j+a_z k) \wedge (b_x i+b_y j+b_z k)$$

When the vectors are parallel, $\phi=0$ so $\sin \phi=0$.

Therefore

$$\bar{A} \wedge \bar{B}=a_x b_y(i \wedge j)+a_y b_x(j \wedge i)+a_x b_z(i \wedge k)+a_z b_x(k \wedge i) \\ +a_y b_z(j \wedge k)+a_z b_y(k \wedge j)$$

Now

$$\bar{i} \wedge \bar{j}=(\sin \pi/2)\bar{k}, \qquad \bar{j} \wedge \bar{i}=(\sin (-\pi/2)\bar{k} \text{ etc.}$$

Hence

$$\bar{A} \wedge \bar{B}=(a_x b_y-a_y b_x)\bar{k}+(a_z b_x-a_x b_z)\bar{j}+(a_y b_z-a_z b_y)\bar{i}$$

i.e.

$$\left.\begin{aligned}(\bar{A} \wedge \bar{B})_x&=a_y b_z-a_z b_y\\(\bar{A} \wedge \bar{B})_y&=a_z b_x-a_x b_z\\(\bar{A} \wedge \bar{B})_z&=a_x b_y-a_y b_x\end{aligned}\right\}$$

Commutation Properties

$$\bar{A}.\bar{B}=\bar{B}.\bar{A}$$

$$\bar{A} \wedge \bar{B}=-\bar{A} \wedge \bar{B}$$

hence

$$\bar{A} \wedge \bar{A}=0$$

Some interesting relationships

(1) $$\bar{A}.(\bar{B} \wedge \bar{C})=(\bar{A} \wedge \bar{B}).\bar{C}$$

PROOF

$$\begin{aligned}\text{L.H.S.}&=a_x(b_y c_z-b_z c_y)+a_y(b_z c_x-b_x c_z)+a_z(b_x c_y-b_y c_x)\\&=(a_x b_y-a_y b_x)c_z+(a_y b_z-a_z b_y)c_x+(a_z b_x-a_x b_z)c_y\\&=\text{R.H.S.}\end{aligned}$$

(2) $$\bar{A} \wedge (\bar{B} \wedge \bar{C})=(\bar{A}.\bar{C})\bar{B}-(\bar{A}.\bar{B})\bar{C}$$

Proof

$$\begin{aligned}\text{L.H.S.} \times \text{component is } & A_y(\bar{B} \wedge \bar{C})_z - A_z(\bar{B} \wedge \bar{C})_y \\ &= A_y(B_xC_y - B_yC_x) - A_z(B_zC_x - B_xC_z) \\ &= (A_yC_y + A_zC_z)B_x - (A_yB_y + A_zB_z)C_x \\ &= (A_yC_y + A_zC_z + A_xC_x)B_x - (A_yB_y + A_zB_z + A_xB_x)C_x \\ &= x \text{ component of R.H.S.}\end{aligned}$$

Similarly for other components.

Formulae involving differentials:

$$\text{del} = \nabla = i\frac{\partial}{\partial x} + j\frac{\partial}{\partial y} + k\frac{\partial}{\partial z} = \text{vector differentiation operator.}$$

Gradient of a scalar function, grad V, is a vector

$$= i\frac{\partial V}{\partial x} + j\frac{\partial V}{\partial y} + k\frac{\partial V}{\partial z}$$

i.e. grad V is a vector whose components are $\partial V/\partial x$, $\partial V/\partial y$, $\partial V/\partial z$. Divergence of a vector is a scalar:

$$\text{div } \bar{A} = \nabla . \bar{A} = \frac{\partial A_x}{\partial x} + \frac{\partial A_y}{\partial y} + \frac{\partial A_z}{\partial z}$$

Curl of a vector is a vector

$$(\text{curl } \bar{A})_x = (\nabla \wedge \bar{A})_x = \frac{\partial A_z}{\partial y} - \frac{\partial A_y}{\partial z} \text{ etc.}$$

The main practical difference between a vector differential operator like ∇ and an ordinary vector is that products no longer commute simply in the general case, i.e.

$$\nabla . \bar{B} \neq \bar{B} . \nabla$$

$$\nabla \wedge A \neq A \wedge \nabla$$

curl grad = 0 for

$$(\nabla \wedge \nabla)_x = \frac{\partial}{\partial y}\frac{\partial}{\partial z} - \frac{\partial}{\partial z}\frac{\partial}{\partial y} = 0$$

$\bar{c} = \text{curl } (\bar{A} \wedge \bar{B}) = \nabla \wedge (\bar{A} \wedge \bar{B})$ for "x" component

$$\begin{aligned} c_x &= \frac{\partial}{\partial y}(\bar{A} \wedge \bar{B})_z - \frac{\partial}{\partial z}(\bar{A} \wedge \bar{B})_y \\ &= \frac{\partial}{\partial y}(A_xB_y - A_yB_x) - \frac{\partial}{\partial z}(A_zB_x - A_xB_z) \\ &= B_y\frac{\partial A_x}{\partial y} + A_x\frac{\partial B_y}{\partial y} - B_x\frac{\partial A_y}{\partial y} - A_y\frac{\partial B_x}{\partial y} - B_x\frac{\partial A_z}{\partial z} - A_z\frac{\partial B_x}{\partial z} \\ &\qquad + B_z\frac{\partial A_x}{\partial z} + A_x\frac{\partial B_z}{\partial z} \\ &= A_x(\nabla . \bar{B}) - B_x(\nabla . \bar{A}) + (\bar{B} . \nabla)A_x - (\bar{A} . \nabla)B_x\end{aligned}$$

i.e. curl $\bar{A} \wedge \bar{B}=\bar{A}(\nabla.\bar{B})-\bar{B}(\nabla.\bar{A})+(\bar{B}.\nabla)\bar{A}-(\bar{A}.\nabla)\bar{B}$

$$\bar{c}=\text{curl curl } \bar{B}=\nabla \wedge (\nabla \wedge \bar{B})$$

$$c_x=\frac{\partial}{\partial y}\left(\frac{\partial B_y}{\partial x}-\frac{\partial B_x}{\partial y}\right)-\frac{\partial}{\partial z}\left(\frac{\partial B_x}{\partial z}-\frac{\partial B_z}{\partial x}\right)$$

$$=\frac{\partial}{\partial x}\frac{\partial B_y}{\partial y}-\frac{\partial^2 B_x}{\partial y^2}-\frac{\partial^2 B_x}{\partial z^2}+\frac{\partial}{\partial x}\frac{\partial B_z}{\partial z}$$

$$=\frac{\partial}{\partial x}(\nabla.\bar{B})-\nabla^2 B_x$$

i.e.

$$\text{curl curl } \bar{B}=\text{grad }(\nabla.\bar{B})-(\nabla.\nabla)\bar{B}$$

i.e.

$$\underline{\nabla \wedge (\nabla \wedge \bar{B})=\nabla(\nabla.\bar{B})-(\nabla.\nabla)\bar{B}}$$

i.e.

$$\underline{\text{curl curl}=\text{grad div}-\nabla^2}$$

Commutation of Operators

Much of the difference between classical and quantum mechanics arises from differences in commutation properties of the variables involved. In classical mechanics dynamical variables can be written as algebraic functions of the space and momentum coordinates. Since in algebra $ab=ba$, the variables all commute with each other and we do not have to be careful about the order in which we write them. As soon as we make the operators corresponding to the variables of a more general form the order does matter.

E.g. suppose

$$f(x)=Ax, \qquad g(y)=By$$

then if

$$fg(x)=f(g(x))$$

then

$$f(gx)=f(Bx)=ABx$$

$$=gf(x)$$

i.e. in this case $fg=gf$, i.e. f and g commute.

On the other hand if

$$h(x)=Ax^2$$

$$hg(x)=h(Bx)=AB^2x^2$$

$$gh(x)=g(Ax^2)=ABx^2\neq hg(x) \qquad \text{(unless } B=1\text{)}$$

for non-trivial cases.

Theorem

If two variables can be observed at the same time then they commute.

Proof: We only observe eigenvalues of variables so if the two operators are F and G then in the state of the system

$$F\,|\,\rangle = f\,|\,\rangle \quad \text{and} \quad G\,|\,\rangle = g\,|\,\rangle$$

where f and g are the observed values. We can label the states $|\,f, g\rangle$.

$$\begin{aligned} FG\,|\,f, g\rangle &= Fg\,|\,f, g\rangle = gF\,|\,f, g\rangle = gf\,|\,f, g\rangle \\ &= GF\,|\,f, g\rangle \end{aligned}$$

i.e. $FG - GF = 0$. Since numbers commute with all functions. The converse is also true. If two operators do not commute they cannot be observed simultaneously—for one of the steps in the preceding proof would then be false, i.e. the state could not be an eigenstate of both operators without involving a logical inconsistency. This is closely connected with Heisenberg's uncertainty principle.

Almost all the properties observed or reflected in magnetic resonance spectra are connected with the lack of commutation between components of the angular momentum. The most useful form of this commutation relation lies in the properties of the "raising" and "lowering" operators. We start from the following:

$$\left.\begin{aligned} M_xM_y - M_yM_x &= i\hbar M_z \\ M_yM_z - M_zM_y &= i\hbar M_x \\ M_zM_x - M_xM_z &= i\hbar M_y \end{aligned}\right\} \qquad \left.\begin{aligned} M_+ &= M_x + iM_y \\ M_- &= M_x - iM_y \end{aligned}\right\}$$

so

$$\begin{aligned} M_zM_+ &= M_zM_x + iM_zM_y \\ M_+M_z &= M_xM_z + iM_yM_z \end{aligned}$$

$$M_zM_+ - M_+M_z = iM_y - i^2M_x = M_+$$

i.e.

$$\underline{M_zM_+ = M_+(M_z + 1)}$$

$$\underline{M_zM_- = M_-(M_z - 1)}$$

$$\begin{aligned} M_+M_- &= M_x^2 + M_y^2 + iM_yM_x - iM_xM_y = M^2 - M_z^2 + M_z \\ M_-M_+ &= M_x^2 + M_y^2 - iM_yM_x + iM_xM_y = M^2 - M_z^2 - M_z \end{aligned}$$

Say $M_+\,|\,l, m\rangle = N\,|\,l, m+1\rangle$ where N is a numerical factor. Then multiply on left by complex conjugate.

$$\begin{aligned} \langle l, m\,|\,M_-M_+\,|\,l, m\rangle = N^2 &= \langle M^2 - M_z^2 - M_z\rangle \\ &= l(l+1) - m(m+1) \end{aligned}$$

i.e.

$$M_+ \mid l, m\rangle = \sqrt{(l-m)(m+l+1)} \mid l, m+1\rangle$$

$$M_- \mid l, m\rangle = \sqrt{(l+m)(l-m+1)} \mid l, m-1\rangle$$

(N.B. if $m=l, M_+=0$

$m=-l, M_-=0$.)

FURTHER READING

On Electricity and Magnetism:

C. A. Coulson, "Electricity", Oliver and Boyd (1948).

B. I. and B. Bleaney, "Electricity and Magnetism", Oxford (1965).

On Quantum Mechanics:

L. Pauling and E. B. Wilson, "Introduction to Quantum Mechanics", McGraw-Hill (1935).

P. A. M. Dirac, "The Principles of Quantum Mechanics", Oxford (1930).

R. P. Feynmann, "Quantum Electrodynamics", Benjamin, (1962).

J. S. Griffith, "The Theory of Transition Metal Ions", Cambridge (1961).

On Magnetic Resonance:

J. D. Roberts, "Nuclear Magnetic Resonance", McGraw-Hill (1959).

D. J. E. Ingram, "Free radicals as studied by Electron Spin Resonance", Butterworths (1958).

P. B. Ayscough, "Electron Spin Resonance in Chemistry", Methuen (1967).

J. A. Pople, W. G. Schneider and H. J. Bernstein, "High Resolution Nuclear Magnetic Resonance", McGraw-Hill (1959).

C. P. Slichter, "Principles of Magnetic Resonance", Harper and Row (1963).

A. Carrington and A. D. McLachlan, "Introduction to Magnetic Resonance", Harper and Row (1967).

A. Abragam, "The Principles of Nuclear Magnetism", Oxford (1961).

The following are the more important papers whose results were used in the discussion/text:

(1) ESR spectrum of oriented $\dot{C}H(CO_2H)_2$ radicals (Ch. 6). H. M. McConnell, C. Keller, T. Cole and R. W. Fessenden, *J. Amer. Chem. Soc.*, **82**, 766 (1960)

(2) ESR of oriented $CH_2(CO_2H)\ \dot{C}H(CO_2H)$ radical (Ch. 4). D. Pooley and D. H. Whiffen, *Molec. Phys.*, **4**, 81 (1961).

(3) NMR coupling constants in simple hydrocarbons (Ch. 6). R. M. Lynden-Bell and N. Sheppard, *Proc. R. Soc.* A **269**, 385 (1962).

(4) ESR coupling constants in aliphatic radicals (Ch. 6). R. W. Fessenden and R. H. Schuler, *J. Chem. Phys.*, **39**, 2147 (1963).

(5) ^{31}P coupling constants in platinum complexes (Ch. 6). A. Pidcock, R. E. Richards and L. M. Venanzi, *J. Chem. Soc.*, 1707 (1966).

(6) Quadrupole effects in the NMR spectrum of pyrrole. J. D. Roberts, *J. Amer. Chem. Soc.*, **78**, 4495 (1956).

(7) CIDNP in photolysis of benzophenone. G. L. and L. E. Closs, *J. Amer. Chem. Soc.*, **91**, 4549 *et seq.* (1969).

(8) NMR spectrum of $(CH_3)_2N\text{–}N=O$. C. E. Looney, W. D. Phillips and E. L. Reilly, *J. Amer. Chem. Soc.*, **79**, 6136 (1957).

Additional useful reviews:

(9) ESR of aromatic radicals. A. Carrington, *Quat. Rev.*, **17**, 67 (1963).

(10) High resolution NMR. J. W. Elmsey, J. Feeney and L. H. Sutcliffe, Vols. I, II Pergamon (1965).

Index